Responsive Landscapes

The sensing, processing, and visualizing that are currently in development within the environment boldly change the ways design and maintenance of landscapes are perceived and conceptualized. This is the first book to rationalize interactive architecture and responsive technologies through the lens of contemporary landscape architectural theory.

Responsive Landscapes frames a comprehensive view of design projects using responsive technologies and their relationship to landscape and environmental space. Divided into six insightful sections, the book frames the projects through the terms: elucidate, compress, displace, connect, ambient, and modify to present and construct a pragmatic framework in which to approach the integration of responsive technologies into landscape architecture.

Complete with international case studies, the book explores the various approaches taken to utilize responsive technologies in current professional practice. This will serve as a reference for professionals and academics looking to push the boundaries of landscape projects and seek inspiration for their design proposals.

Bradley Cantrell is an Associate Professor at the Harvard Graduate School of Design, USA, and design researcher at Invivia whose work focuses on the role of computation and media in environmental and ecological design. Professor Cantrell received his BSLA from the University of Kentucky and his MLA from the Harvard Graduate School of Design. He is the author of the ASLA award-winning book *Digital Drawing for Landscape Architecture and Modeling the Environment.*

Justine Holzman is a landscape researcher and adjunct Assistant Professor at the School of Landscape Architecture at the University of Tennessee, Knoxville, USA. She received a BA in Landscape Architecture from the University of California Berkeley and an MLA from the Robert Reich School of Landscape Architecture. Her current research recognizes the inherent responsive capabilities of landscape through its materiality and her work in ceramics and digital fabrication has been exhibited across the United States.

Responsive Landscapes

STRATEGIES FOR RESPONSIVE TECHNOLOGIES IN LANDSCAPE ARCHITECTURE

Bradley Cantrell and
Justine Holzman

LONDON AND NEW YORK

First published 2016
by Routledge
2 Park Square, Milton Park, Abingdon, Oxon OX14 4RN

and by Routledge
711 Third Avenue, New York, NY 10017

Routledge is an imprint of the Taylor & Francis Group, an informa business

British Library Cataloguing-in-Publication Data
A catalogue record for this book is available from the British Library

Library of Congress Cataloging-in-Publication Data
Cantrell, Bradley, author.
Responsive landscapes: strategies for responsive technologies in landscape architecture/Bradley Cantrell and Justine Holzman. –
First edition.
pages cm
Includes bibliographical references and index.
1. Landscape architecture. I. Holzman, Justine, author.
II. Title.
SB472.C252 2016
712′.5–dc23 2015018404

ISBN: 978-1-138-79665-2 (pbk)
ISBN: 978-1-315-75773-5 (ebk)

Typeset in Avenir
by Florence Production Ltd, Stoodleigh, Devon, UK

I would like to give personal thanks to my family, Susan, David, and Hannah. Thank you.—Brad

I would like to thank my parents Allan and Susan, whose lives are a creative endeavor, and my sister Shayne, who is a constant inspiration.—Justine

CONTENTS

FIGURES

FOREWORD

Towards a robotic ecology

Jason Kelly Johnson and Nataly Gattegno
Future Cities Lab

Sensing, processing, visualizing, and feedback: These are the key processes that this volume hypothesizes are a new conceptual methodology for landscape and ecology in the coming era. Each process, referencing predominantly technical disciplines, suggests that the domain of emerging landscape practices will increasingly cross-over into fields such as computer science and robotics. In some ways this new methodology simply builds upon well-established disciplinary topics such as time, phasing, and entropy; however, in other ways this volume suggests something more radical: it forecasts an emerging world of *robotic ecologies*, where matter at all scales is programmable, parametric, networked, and laden with artificial intelligence.

Responsive Landscapes engages a latent territory that, to date, has remained largely underexplored within the discipline of Landscape Architecture. Authors Cantrell and Holzman predict an emerging paradigm shift—where biology, intelligent machines, and systems will begin to productively co-exist and co-evolve. By coupling this synthetic shift with the ubiquity of networked technologies and open-source resources, tomorrow's designers will be able to explore, design, and construct landscape prototypes that have in the past remained unapproachable. By experimenting across scales, for instance linking sensor-laden physical models to much larger and complex ecological simulations, the potential impact of these methodologies on landscape and infrastructure scale explorations is highly promising.

The authors argue for a conceptual shift from a more object-oriented understanding of technology as a mediator between systems to a more integrated and synthetic understanding of technology as the medium through which we can encode and amplify landscapes with intelligence and heuristic capacities. In other words, when landscapes get hybridized with responsive technologies, they will have the capacity to better process and respond to the variable and multi-scalar inputs from their environments. As the collected projects in this volume suggest, these sensing inputs and cybernetic capacities are now possible mediums to be experimentally mined for their spatial, material, and ecological potentials.

What would typically be described as 'form' is, in this methodology, defined as 'output': physical and digital manifestations perform in parallel and exchange real-time information across scales, contexts, and project types. The cross-disciplinary projects included in this volume provide a tantalizing sense of the texture and palette these outputs can produce. Far from passive visualizations of complex information, these tangible data-informed landscapes create visceral, immersive, and participatory human experiences. In the past, to allow for the sheer processing of large datasets, inputs were limited to two or three select parameters. With our increased capabilities to process large amounts of information, the authors believe that parallel models and simulations—what they call "hybridized" models operating simultaneously—could prove to be more effective models at the scale of landscape. Rather than argue for one large data processing model that encompasses climate change, hydrology, geology, flora, fauna, etc., the authors call for parallel virtual models that run concurrently to create a hybrid feedback loop for the evaluation of multiple possible design trajectories in real-time.

The virtualization and networking of the physical world also opens up the opportunity for communication and feedback with systems such as social networks and the "internet of things." At the scale of landscape, one could argue that "feedback" has typically been explored as a linear relationship between a landscape and its built environment: data about biological processes accumulated over long periods of time and in a more or less static context. What are the implications when these feedback loops are real-time, when contextual data is dynamic, and our algorithms allow us to evolve life-like characteristics and robotic ecologies? Several projects in this volume explore the opportunities for encoding behaviors such as unpredictability and randomness into landscape methods. By reconceptualizing the duration of landscape, this volume also posits that the concept of time must also be rethought. How do systems that we usually consider operating at the scale of days, months, years, seasons, or centuries respond to instantaneous dynamic inputs of information?

While the opportunities for this methodology are tremendous, they are clearly still being formed. Through essays and projects that cross scales and disciplines, the authors map out latent territories for landscape architecture to explore and give form in the coming years. Using terms such as "elucidate," "compress," "displace," "connect," "ambient," and "modify," they suggest a developing lexicon and conceptual framework ripe with possibilities. This raises questions such as: How will this framework transform the way we understand and design the robotic ecologies of the future? How will we construct and encode them? How will we interact with them? How will we navigate through them? And, perhaps the most challenging question, what pedagogical approaches will the discipline need to adopt to explore and engage these responsive landscapes to their fullest extent? Cantrell and Holzman, in the essays and projects collected here, suggest a methodology for the next generation of landscape designers. They call on students, designers, and educators to take ownership of these emerging methodologies and, most importantly, to engage them as design opportunities with mounting social, political, and ecological implications.

ACKNOWLEDGMENTS

This book would not be possible without the support of colleagues who provided critical feedback, students that displayed their unbound creativity, and the support of our institutions. Louisiana State University has provided a progressive platform to expand the discourse of responsive technologies through funded research in the LSU Coastal Sustainability Studio and in option studios at the Robert Reich School of Landscape Architecture. The Harvard Graduate School of Design has provided a platform through the Responsive Environments and Artifacts Lab and the MDes Technology concentration to imagine the role of landscape in responsive systems. The book relies heavily on the contributions designers have made through the case studies and interviews; so we would like to say thank you for allowing us to highlight your amazing work and understand your methodologies. We would also like to specially thank the contributions of Allen Sayegh, Jason Kelly Johnson, Nataly Gattegno, and Nicholas de Monchaux who went out of their way through unique contributions and mentorship. And to Rob Holmes for his early notes on the formation of the Manuscript.

Figure 01.01 Ecolibrium, *Kim Nguyen, Devin Boutte, Martin Moser, Joshua Brooks, Responsive Systems Studio, 2011*

INTRODUCTION

01

THE PARADIGM SHIFT

The last two decades have seen a range of experiments using responsive technologies focused on the interaction between environmental phenomena and architectural space. These experiments go beyond site or architectural controls that rely on efficiency and automation instead they are attempts to expand the application of responsive technologies. Novel and explorative work within this realm has emerged as installations or unique architectural features, often requiring collaborations across disciplinary boundaries and the hacking of accessible technologies. This text highlights a collection of projects experimenting with the application of responsive technologies and pulls forth methods specifically related to the indeterminacy and dynamics in contemporary landscape architecture. The application of responsive technologies in architecture has become technically advanced, but is ". . . in fact responding to the question posed in the 1960s by Cedric Price: What if a building or space could be constantly generated and regenerated?"[1] For landscape architects the act of response and regeneration is the basis of our profession and inherent to landscape as a medium. Therefore it is necessary to understand a framework for responsive technologies that speaks to the scale of the territory and acknowledges the interconnections of the many.

The advancement and availability of responsive technologies have increased accessibility to designers, prompting the development of new design methodologies that move beyond conventional methods of representation and implementation. The introduction of accessible software sets the stage for design culture to appropriate and advance software and hardware tools.[2] New methods focus on the expression or design of processes, logics, and protocols requiring design interventions to evolve throughout a project's lifespan. Evidenced by Usman Haque and Adam Somlai-Fischer's open-source research report, "Low Tech Sensors and Actuators for Artists and Architects,"[3] detailing the hacking and re-purposing of low-cost and widely available technologies embedded in toys and standard devices as a method for artists, architects, and designers to quickly and effectively prototype responsive and interactive urban installations that would otherwise require client support. In a similar manner ". . . during the 1980s GUI-based software quickly put the computer in the center of culture,"[4] the advent of visual programming is putting coding and scripting directly in the hands of designers. The coupling of Arduino IDE boards and kit-of-parts beginner robotic kits with software plugins to easily program unique methods of response have further hastened the pace of artists and designers prototyping innovative interactive solutions to urban scale problems.

Landscape architecture has seen a paradigm shift in the last two decades, requiring designers to respond to the dynamic and temporal qualities of landscape. This response examines the long-held view that landscape embraces an ephemeral medium constructed and maintained through generations. Landscape—a dynamic and temporal medium—is expressed through careful manipulation of vegetated, hydrological, and stratigraphic systems. Combining this shift with the increased accessibility of responsive technologies presents a new approach for challenging static design solutions. The ability to sense and respond to environmental phenomena invites new ways to understand, interpret, experience, and interact with the landscape.

This shift can be traced to several parallel events inherent to the discipline of Landscape Architecture and seeded by new paradigms in scientific thought particularly within ecology. A generational trend has emerged within landscape architecture that promotes a form of "distanced authorship,"[5] emphasizing natural processes such as succession, accretion, or passive remediation as agents for landscape design. In the essay, "Strategies of Indeterminacy in Recent Landscape Practice," Charles Waldheim uses the term "distanced authorship" to describe how the "privileging of landscape strategy and ecological process distances authorial control over urban form, while allowing for specificity and responsiveness to market conditions as well as the moral high-ground and rhetorical clarity of environmental determinism."[6] Autonomy within these systems has the potential to create scaffolds for designed landscapes, urbanism, or territorialization. This approach privileges the actions of biology and geology over manufactured static conditions and instead seeds these dynamic processes through an overarching ecological regime to shape designed conditions over time.

In the introduction to *Case: Downsview Park Toronto*, Julia Czerniak synthesizes this shift, traced from the international design competition for Parc de la Villette (1982/1983), towards "process" and "ecological frameworks," . . . reshaping landscape perceptions to value "processes of becoming," "frameworks over form," and performance.[7]

Bernard Tschumi's team proposal frames processes around a few key species and relies on processes of succession to build complexity over time, creating a known starting point and a maintenance regime that embraces flux. James Corner and Stan Allen's team proposal, titled "Emergent Ecologies," engages the concept of *emergence* as the combination of intentional and unintentional futures shaped by ecology and human intervention as an "engineered matrix" performing as a "living groundwork for new forms and combinations of life to emerge."[8] Corner and Allen boldly state, "we do not determine or predict outcomes; we simply guide or steer flows of matter and information."[9]

Continuing along this trajectory, in 2002 Field Operation's proposal for Fresh Kills in Staten Island highlighted phasing and indeterminacy as central agents in design. Fresh Kills is a brownfield landscape of significant scale requiring novel methods for performative uses of vegetation with minimal maintenance regimes. This approach bridges earlier projects redefining the discipline of Landscape Architecture that focused on post-industrial remediation, to expand the scope, scale, and potential for remediation and evolving landscapes. Field Operations uses a similar method of seeding vegetation within bands tied to the elevations of the landforms (landfills).

What emerges from the late 1990s in landscape architecture is over two decades of exploration that has focused on complexity, indeterminacy, and dynamic systems. This body of research is marked by texts such as *The Landscape Urbanism Reader*[10] edited by Charles Waldheim (2006); *Ecological Urbanism*[11] edited by Mohsen Mostafavi and Gareth Doherty (2010), key categories of which are "sense," "curate," "interact," and "measure"; and most recently *Projective Ecologies*[12] edited by Nina-Marie Lister and Chris Reed (2013), which draws together a reader of seminal essays contributing to this discourse around concepts of "dynamics," "succession," "emergence," and "adaptability." This direction for the discipline continues to evolve the concept of "distanced authorship"[13] through a series of practices that have fought to realize built works. Landscape Architecture is a discipline of making. Practitioners and academics have sought to employ a multitude of techniques to understand how landscapes evolve and interrelate. On one hand, the profession has engaged and developed workflow methodologies with state-of-the-art tools in computation to simulate, analyze, and spatialize huge datasets to understand complex ecological relationships. On the other, landscape architects have pushed this agenda through the traditional tools of drawing, modeling, and diagramming to describe these complex systems, essentially outlining the projective tools they need. At this moment, there are trajectories for new computational methods beginning to find traction tied to a lineage of representational methods interrogating time through drawing and photographic methods such as the static series, image sequence, and photographic recording methods. This mode of seeing and transforming through an increased faculty with computational tools brings forth a new project for landscape that is firmly seated in an evolving ecological framework—a framework which, through distanced authorship, intends to address landscape of larger scales with more complex ecological problems tied to settlement and industry.

An ecological framework for landscape architecture is one that is based on strategy, an approach to landscape inextricably tied to habitat, species, and culture. Kate Orff describes that her "intuitive leap towards landscape begins with imagining the life it carries: mammals, molluscs, protoplasm" when describing her re-reading of Rachel Carson's 1937 book, *Undersea*, for *Harvard Design Magazine.*[14] This attachment to ecology through the species and individuals is a relationship that landscape architects and other environmentally based disciplines state as inspiration. It is also a powerful mechanism that pulls the public into ecologically based projects. This sentiment, coupled with advances in ecological sciences and a mandate for landscape architectural practice to adopt a strategic mandate, is the framework landscape architects rely upon.[15] This evolving framework is perfectly suited as a basis for utilizing responsive technologies and computation in ecological systems.

The ability to implement new computational methodologies hinge around emerging technologies for sensing and responding to real-time conditions. Responsive technologies counter disturbances through self-regulating systems, apparent when, "the linear system disturbs the relation the self-regulating system was set up to maintain with its environment."[16] Responsive technologies play a pivotal role in our evolving relationship between constructed and evolved systems. Current models of machine/human interaction are quickly evolving to encompass more complex methods of simulated intelligence and nuanced response. Several technologies that change the landscape of responsive technologies are converging, including autonomous robotics, distributed intelligence, biotic/abiotic interfaces, and ubiquitous sensing networks. As early as the 1980s, Xerox PARC coined the term "ubiquitous computing," which imagined the evolution of the human computer interface to "[take] into account the natural and human environment and [allow] the computer to vanish into the background."[17] With this focus away from HCI as personal device and integration into the environment, these technologies

Figure 01.02 Synthetic territories diagram, Bradley Cantrell, 2011

fundamentally alter our perception of constructed systems and their nuanced relationships with ecological processes.

These technologies have been recognized within architecture for their potential to create flexible and adaptable (though not adaptive in the ways ecological systems have the capacity to evolve) spatial or social conditions. "While, arguably, architecture has always been responsive, encouraging interaction between a space and the people that use it, new technological developments are putting pressure on architecture to become more adaptable and intelligent."[18] The extent to which responsive technologies address the goals of contemporary landscape architectural theory remains an emerging field. *Responsive Landscapes* conceptualizes the connection between environmental phenomena and responsive technologies as a continuum in which landscape places a vital role. The sensing, processing, and visualizing we are currently developing within the environment boldly changes the ways we perceive and conceptualize the design and maintenance of landscape or environment. Both *Interactive Architecture*[19] by Michael Fox and Miles Kemp (2009), and *Responsive Environments*[20] by Lucy Bullivant (2006) have set precedents for the integration of responsive technologies in the field of architecture. *Interactive Architecture* highlights malleable systems and transformable morphologies, whereas *Responsive Environments* begins to point towards more nuanced relationships between architectural objects as mediators of space and interaction. *Responsive Landscapes* is the first work that attempts to rationalize interactive architecture and responsive technologies through the lens of contemporary landscape architectural theory. These new relationships suggest a series of networked and object-oriented relationships between designed devices, ecological entities, and regional influences. This shift calls for an expanded view that asks for ecological system abstraction, filtering, and embedded intelligence that drives feedback loops of sensing, processing, and visualizing. This process of feedback, sensing the environment, processing the sensed data, and visualizing the response is the core design focus in the development of responsive technologies.

NATURE?

A fundamental aspect to further understanding the role of responsive technologies as drivers of landscape scale manipulations is the often dualistic view of human/nature interactions that has shaped the discipline of Landscape Architecture. Our relationship with the natural environment can never be described simply. This dualism of clearly delineating objects and processes within the world as a product of nature or as a product of humanity has created a perceived separation of interaction. Over several decades, new understandings of ecology tied to ecological disturbance make it overtly apparent that we live in an environment constantly evolving in parallel to our interactions with it. While not under our control—these environments are synthetic expressions of both direct and indirect anthropogenic interaction with environmental processes. As the discipline attempts to shift formative conceptions of human/nature interactions and operate within an anthropogenic biosphere, designers are drawing from new definitions and re-conceptions of ecology, ecological thought, and geologic scale change from multiple disciplines including philosophy and the sciences.

Linda Weintraub's definition of "deep ecology":

> . . . [a] philosophy that envisions the universe as unified and interconnected and recognizes the inherent worth of all forms of life without regard for human utility and pleasure. As such, deep ecologists pursue metaphysical unification of humans and their surrounds, as opposed to relying on reason, to guide environmental reform.[21]

Understanding the environment human beings operate in, as a composite product of our interactions and a series of systems, allows for designers to operate as active agents within an assemblage of biotic and abiotic agents. As designers we can understand our role differently—if we are no longer in opposition to the operation of ecological systems we can assume the roles of curators and manipulators of processes.[22] Within this new mode of operation, designers are using and developing new tools to understand historic processes and future outcomes while working within a localized environment.

The environment we operate within can be seen as an anthropogenic product, where human beings are one of many contributors within an ecological system. While our scope is wider and our effects more prolific, our modes of construction and habitation are an integral (although at times disruptive) portion of the ecological systems in which we are situated.[23] Evidence of a new geologic period is easily found in the altered stratigraphy of cities, rapid population growth in response to synthetic nitrogen production, the homogenization of biodiversity across the globe by the domestication of plants and animals, mass species extinctions, and dramatic increases in atmospheric carbon. Ellis and Ramankutty identify 18 anthropogenic biomes through empirical analysis of global population, land use, and land cover, that reside outside of existing descriptions and representations of biome[24] systems as "ignor[ing] humans altogether or simplify[ing] human influence into, at most, four categories."[25] Their research offers a way to assess current conditions of the terrestrial biosphere by providing accurate models depicting the true immersion of human and ecological systems. Anthropogenic biomes elucidate a relationship defined by human systems with natural systems embedded within them.

The emerging philosophical fields of new materialism and object-oriented-ontology, are useful for situating the designer's role as curator or manipulator of processes—considering both biotic and abiotic factors as equally engaged in shaping environments. Jane Bennett, a new materialist and author of *Vibrant Matter: A Political Ontology of Things*, elaborates on a further hindrance to building an effective view of contemporary ecological systems predicated on the false assumption that non-human matter is inanimate—though arguably non-human agency is required for human intent and intervention to manifest—and considers the capacity of things as equal actants.[26] Bennett uses materiality as "a rubric . . . to horizontalize the relations between humans, biota and abiota," indicative of the potential for responsive procedures within the landscape to actively shape material driven landscape processes.[27] Speaking to the political capacity of agentic assemblages, she uses the example of worms, free to make unpredictable decisions in the face of different material situations given different types of soils and ground covers, that ultimately contribute to a larger ecosystem responding in real-time without an overall goal or pre-determined outcome. In this example, materials play a vital role in the function, performance, and shifting configurations of ecosystems—such that, "the figure of an intrinsically inanimate matter may be one of the impediments to the emergence of more ecological and more materially sustainable modes of production and consumption."[28]

Both new materialism and object-oriented ontology (though unique fields of philosophical thought) provide ways into process based approaches to landscape manipulations beyond human intentionality. The approaches aim to attach the manipulation of landscape over time to the importance of site specificity—design should be based on unique phenomena of location and site history. The current state of a landscape is not the final state; rather it is a moment within a larger history and context as site processes are ongoing. Thus, an ecological state is not defined by a pre- or post-condition, but is continuously acting and evolving. Site-specific sensed data can provide curated histories over time to extract knowledge of material-based processes in order to inform future histories. This approach allows for movement between scales of time and space, to identify processes associated with ecological imperatives.

The effect of human inhabitation on the planet can be seen through different settlement and infrastructure cases. A common case is the infrastructural systems of the Mississippi River watershed, composed of over 80 rivers and tributaries. The watershed interfaces a myriad of urban, agricultural, logistic, and cultural sites from Lake Itasca in the north, the Allegheny River in the east and the Milk and Missouri rivers in the west. This sprawling watershed covers over 40 percent of the continental United States, approximately 1.2 million square miles with the Mississippi River itself running for nearly 2,550 miles.[29] It is within this massive region that human beings have slowly manipulated and altered the course, speed, and scope of the river system in an attempt to provide consistent outcomes for navigation and the protection of urban centers. These manipulations have taken the form of large-scale infrastructure such as the Old River Control Structure, the Bonnet Carré Spillway, and the extensive levee systems that channel the river and prevent the flooding of the landscapes outside of the river's main channel. There are also small-scale manipulations such as channel dredging, vegetation removal, and bank stabilization that are reoccurring and help to promote stability within this landscape.

The management of this river system provides a remarkable landscape for human settlement or commerce and focuses primarily on the current condition with little regard for future scenarios.[30] The Mississippi River watershed used to function as a meandering water system, engaging a myriad of systems from the micro to the macro scales. The argument could even be made that this river system affects climate at a global scale and genetics at the molecular level, thus touching every level of earthbound relationships. This relationship is not only dynamic, but it is also symbiotic: as the river channel breaks, territory is flooded, mixing upstream sediments with latent backwaters. This ever-evolving relationship between local systems and continental shifts is the key to the health of our ecological systems. The Earth does not survive through sequestered zones where we quarantine systems and processes to produce singular desires.

A clearer example, as an engineered system in contrast with naturalized systems, is the Los Angeles River. The current form of the channel no longer resembles our understanding of the definition of "river." Our perception that rivers are the product of an upstream water source that is attempting to find lower ground is skewed as we are confronted with a looming concrete channel with very little water. There are no meandering waterways punctuated with clusters of plant material, seasonally filling and washing through the basin. Instead the river has been engineered to control an annual torrent of water that rushes down from the mountains and fills the channel before hastily rushing out to the mouth of the river in Long Beach. It is often quoted as being dead, devoid of the life that defines the geographic typology of "river."

This engineered solution is a marvel, providing a stability to downtown Los Angeles that was missing from a river that used to be wildly unpredictable.[31] Similar to other engineering solutions during this time period, the solution works on a single goal: to move water at near constant velocities from north of the city to the south during times of high rain or snow melt. This adherence to a single goal has pushed aside the multiplicity of systems like the Los Angeles River, forgetting that these are habitats, methods of recharging aquifers, cleansers, retreats, and many other imagined and unimagined forms that engage our world. This is a story that plays out across the globe as human beings engage with the world around them.

The greatest fault in these engineered systems is the lack of engagement with complexity, particularly the processes feeding into and extending from them. While it is easy to criticize these systems, they must be understood as novel landscapes, providing a key translation of the relationship between synthetic constructions and endemic ecologies. These indirect products of landscape intervention can often be seen as one-offs or, as David Fletcher describes the Los Angeles River, a "freakology," a way of defining the "churning soup of exotic and native vegetative communities" despite the highly industrialized and contaminated urban condition of the channelized river.[32] This is often a central criticism but these are apt descriptions that frame the product of the systems as something different, unusual, or even unnatural. Rather than difference, or uniqueness, the criticism should lay in the kitsch, or lack of complexity and heightened biological stress that typically parallels these moments. These novel relationships are important, particularly in an approach that creates a new series of connections to contextual systems as we are creating systems that are new and unique for their context.

While humans may have little direct access to certain environments, we are still within their sphere of influence. Landscape architecture has always inserted itself within the relationship between humans and "nature," but the implications of acknowledging the Anthropocene as the current geologic epoch for landscape architecture and design of the built environment offers a much more accurate depiction of the scope and potential for intervention within anthropogenic biomes. Studies reveal that human-dominated ecosystems make up a higher percentage of the Earth's terrestrial area in comparison to still "wild" ecosystems. Historical ecologists, proponents of an interdisciplinary research program rooted in cultural anthropology, argue that humans

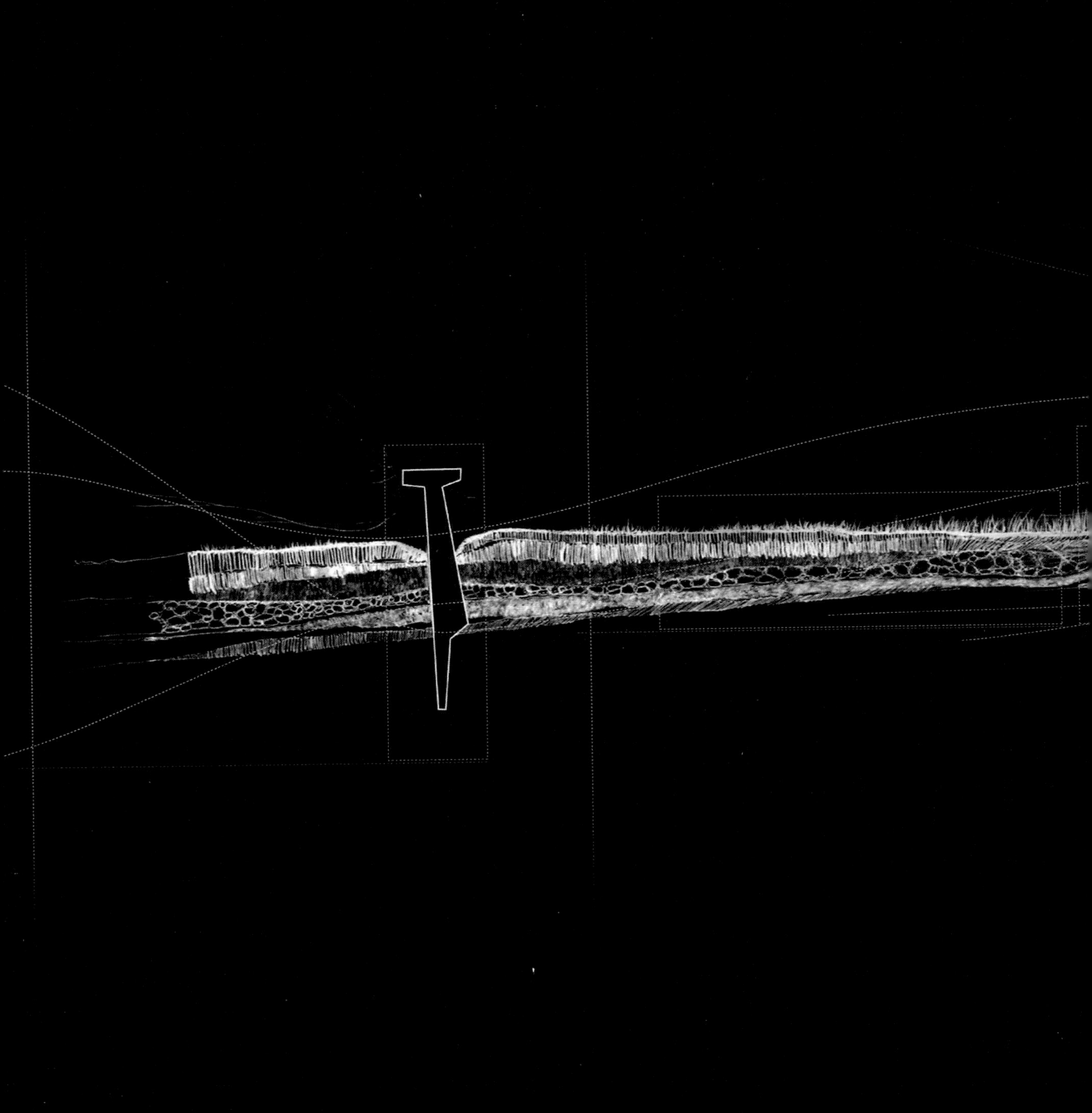

Figure 01.03 Cyborg Landscapes, *Bradley Cantrell, Kristi Cheramie, and Jeffrey Carney, 2010*

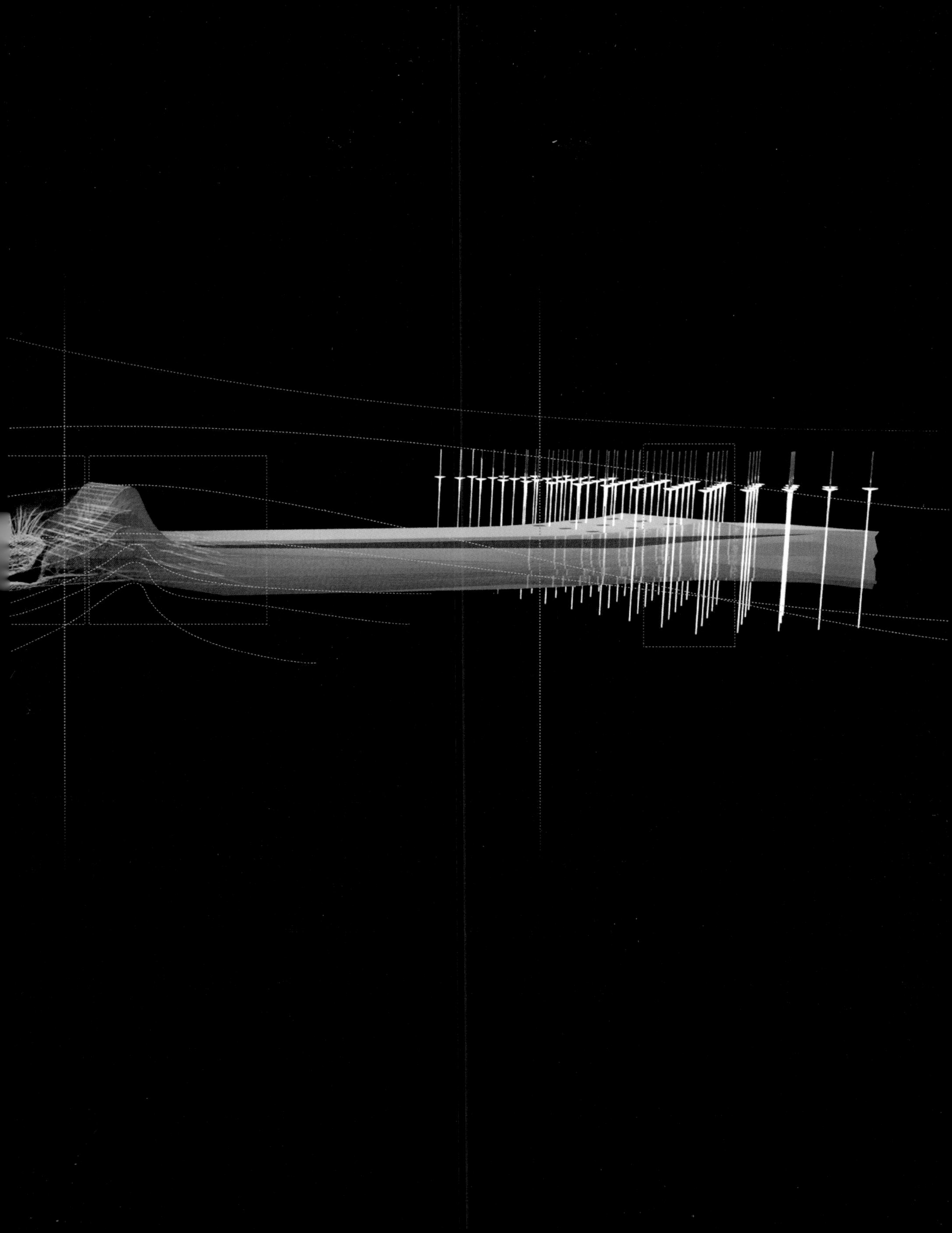

have shaped nearly every environment on Earth.[33] William Balée, historical ecology's primary advocate, notes even the supposed "pristine" forests of Amazonia encode a deep horticultural past in the presence of accessional species and in patches of Amazonian Dark Earth (ADE).[34]

Reconceptualizing our relationship with natural systems, our priorities must encompass human needs while balancing ecological fitness. To do this it requires a very difficult cognitive shift that does not separate human actions from non-human processes. We are required to confront complexity and diversity as driving principles in our relationships with natural systems and to develop equally complex computational relationships. It is through this lens, that we decenter humanity and clarify that we are not separate from "nature" but instead that we are "nature."[35] Speculating on the design methodologies and frameworks of responsive systems provides requires that we develop negotiations rather than controls. These negotiations require that humans are not placed in a mythical place of privilege but encapsulate the idea that concepts of ecological fitness and human comfort are equally considered.[36]

The device, machine, or object inherently influences the field in which it acts or resides. This form of instrumentality varies but can be framed as the ability of an object to influence the development of much larger and complex systems. Anthony Burke describes this networked condition as "Spatially indeterminate, temporally contingent, unstable, inclusive, and dynamic," in which network exhibits the "condition of paradoxical inclusion more aligned to quantum mechanics than the either/or of a discursive modernism."[37] This plays out in several different ways, as emergent behavior, interrupters, or through co-evolution. Instrumentality through emergence can be seen in a variety of biological systems, evolving from small changes in ecological systems that propagate into larger changes in systems. An example may be illustrated through the processes of commercial fishing and understanding the tools of this process as landscape instruments. These tools, fashioned to extract specific species of aquatic life, remove actors (fish, crabs, shellfish) from a large system and catalyze events that create changes along the food chain, within the water column, and may even affect climate. In a similar manner it is possible to describe interrupters as landscape instruments; this may be something similar to a dam, windbreak, or highway constructed to create a static condition within the landscape. This interruption in a larger system creates a series of effects that alter environments.

LOOKING FORWARD

This confluence of rapid technological development, an expanded view of humanity and the environment, and the influence of anthropogenic processes creates a tenuous state that requires an important shift in our conceptualization of responsive technologies. The landscapes that we can begin to imagine have the capacity to not only embed themselves within their context, but can also evolve with a life of their own, a synthesis between the biological, mechanical, and computational. There are several aspects that must be addressed in this regard, particularly in reference to our relationship to the design of systems that focus predominantly on control. Adam Greenfield asks:

> How might we use networked technologies to further the prerogatives so notably absent from the smart-city paradigm, particularly those having to do with solidarity, mutuality and collective action?[38]

This question, directed at how we design our cities, is our mandate for landscape systems at the city and territorial scales. In order to unpack this notion, it is critical that we understand the nature of our interactions: are they about embedding new forms of intelligence or do they simply imply a tightening of a feigned control over chaotic systems? This is a topic that is addressed not only in how we model and visualize these systems but also in how we use this information to create methods of maintenance, construction, and evolution. This requires a view of the devices and infrastructures that are implemented and also the communications and interactions that occur between these systems. "Following from Watts and Strogatz, a protocological architecture necessarily exists in the in-between space, the topological fold of both an empowering infrastructural ambience, and points of concentration that effectively organize that ambience."[39] This liminal space—or more aptly, the landscape itself—becomes the area of concern.

These forms of embedded intelligence must be confronted across scale and time, which are drastically shifted within this new paradigm. Scale is not only about relationships spatially, from site to territory, but also refers to the extents of the issues that designers are confronting at the global scale. The ability to address problems at the global scale requires more than monumental physical engineering; it requires a deft and evolving set of methods that fully adopt the complexity of ecological relationships. The scale of these issues also exists within a new temporal space. This space asks that landscapes are responsive to local processes as well as geologic shifts. These two scales of time were once out of our reach, but are slowly becoming clearer through simulations and models.

The projects outlined in this text are classified by their potentials for responsive design within the landscape and their effects, either perceptually or by direct manipulation. The case studies open up a focused discussion, framing the potentials for responsive technologies, that takes the installations beyond the role of landscape or architectural folly. The structure of the text responds to several modes that are specific to landscape methodologies, and organizes the case studies into actions expressed by the responsive system. A common series of threads are found through the projects, speaking to their direct relationship to the landscape by expressing each project's mode of response. These actions recognize the modes of behavior or modifications that phenomena are subjected to when converted to data or expressed through analog constructions. These responses focus on methods of clarity, interaction, connectivity, and augmentation that deliver both physical and virtual environments, expanding beyond installations to new relationships within the landscape. In the Foreword of *Alive: Advancements in adaptive architecture*, Carole Collet writes:

> In times where the very concept of 'nature' is questioned not only in its philosophical dimension, but in the core of its biological materiality, we need to reconsider the interrelations between architecture and nature.[40]

Moving forward from this cognitive shift, *Responsive Landscapes* frames a comprehensive view of interactive or responsive projects and their relationship to landscape or environmental space. *Responsive Landscapes* deconstructs a series of contemporary projects to develop a lexicon that defines new methods of constructing and framing responsive systems. Many of the projects are speculative and demonstrate a new methodology of working that moves beyond conventional methods of representation or perception. The complexity embedded in the design of responsive technologies requires iterative prototyping and computational development. This process of prototyping requires rigorous methods of making to tune sensing, feedback, and actuation. Each of the projects in *Responsive Landscapes* engages feedback or response as a method of modification, in a limited way, to understand the environment and to respond in calculated ways. While many of the selected projects are not specifically "landscapes," each engages landscape in important ways and develops a pragmatic framework to understand responsive methods in a new context.

NOTES

1. Lucy Bullivant, "Introduction," *Architectural Design* 75, no. 1 (2005): 5.
2. Lev Manovich, *Software Takes Command*. Vol. 5, *International Texts in Critical Media Aesthetics Founding*, edited by Francisco J. Ricardo (New York, NY: Bloomsbury, 2013), 21.
3. Usman Haque and Adam Somlai-Fischer, "Low Tech Sensors and Actuators for Artists and Architects," in *Research: The Itemisation of Creative Knowledge*, Ed. Clive Gillman (Liverpool, UK: FACT: Liverpool University Press).
4. Manovich, *Software Takes Command*, 21.
5. Charles Waldheim, "Strategies of Indeterminacy in Recent Landscape Practice," *Public* 33 (Spring 2006): 80–86.
6. Ibid.
7. Julia Czerniak, Ed., *CASE: Downsview Park Toronto* (Munich: Prestel, 2001), 13.
8. James Corner and Stan Allen's project description for their proposal "Emergent Ecologies" in Czerniak, *CASE—Downsview Park Toronto*, 58.
9. Ibid.
10. Charles Waldheim, Ed., *The Landscape Urbanism Reader* (New York: Princeton Architectural Press, 2006).
11. Mohsen Mostafavi and Gareth Doherty, Eds., *Ecological Urbanism* (Baden, Sweden: Lars Müller, 2010).
12. Chris Reed and Nina-Marie Lister, "Introduction: Ecological Thinking, Design Practices," in *Projective Ecologies*, 14–21.
13. Waldheim, "Strategies of Indeterminacy in Recent Landscape Practice."
14. Kate Orff, "Re-Reading: Rachel Carson, 'Undersea' (1937)," *Harvard Design Magazine* 39, (Fall/Winter 2014): 115.
15. James Corner, "Not Unlike Life Itself: Landscape Strategy Now," *Harvard Design Magazine* 21 (Fall/Winter 2004): 32–34.
16. Hugh Dubberly, Usman Haque, and Paul Pangaro, "ON MODELING: What is interaction?: are there different types?" *Interactions* 16, no. 1 (January 2009): 69–75.
17. Mark Weiser, "The Computer for the 21st Century," *Scientific American* 265, no. 3 (1991): 94–104.
18. J. Meejin Yoon and Eric Höweler, *Expanded practice: Höweler + Yoon Architecture/My Studio* (New York: Papress, 2009), 9.
19. Michael Fox and Miles Kemp, *Interactive Architecture* (New York: Princeton Architectural Press, 2009).
20. Lucy Bullivant, *Responsive Environments: Architecture, Art and Design* (London: V & A Publications, 2006).
21. Linda Weintraub, *To Life!: Eco Art in Pursuit of a Sustainable Planet*, (Berkeley, CA: University of California Press, 2012), xxxv.
22. Chris Reed, "The Agency of Ecology," in *Ecological Urbanism*, Eds. Mohsen Mostafavi and Gareth Doherty (Baden Sweden: Lars Müller, 2010).
23. "The Anthropocene: A man-made world," *The Economist*, May 26, 2011.
24. Erle. C. Ellis, and Navin Ramankutty, "Putting People in the Map: Anthropogenic Biomes of the World," *Frontiers in Ecology and the Environment 6*, no. 8 (2008): 439–447.
25. Ibid.
26. Jane Bennett, *Vibrant Matter: A Political Ecology of Things* (London: Duke University Press, 2010), 9.

27 Ibid., 112.

28 Ibid., ix.

29 J.C. Kammerer, "Largest Rivers in the United States," *U.S. Geological Survey*, May 1990, http://pubs.usgs.gov/of/1987/ofr87-242/ February 22, 2011.

30 Jim Addison, "History of the Mississippi River and Tributaries Project," US Army Corps of Engineers New Orleans District, archived from the original on January 28, 2006, http://wayback.archive.org/web/20060128111022/www.mvn.usace.army.mil/pao/bro/misstrib.htm July, 2015.

31 Blake Gumprecht, *The Los Angeles River: Its Life, Death, and Possible Rebirth* (Baltimore: Johns Hopkins University Press, 1999).

32 David Fletcher, "Flood Control Freakology: Los Angeles River Watershed," in *The Infrastructural City: Networked Ecologies in Los Angeles*, Ed. Kazys Varnelis (New York: Actar, 2009), 42.

33 William Balée, "The Research Program of Historical Ecology," *Annual Review of Anthropology*, no. 35 (2006): 76.

34 William Balée, "People of the Fallow: a Historical Ecology of Foraging in Lowland South America," *Conservation of Neotropical Forests: Working from Traditional Resource Use*, Eds. Kent H. Redford and Christine Padoch (New York: Colombia University Press, 1992), 35–57.

35 Kristina Hill, "Shifting Sites," in *Site Matters: Design Concepts, Histories, and Strategies*, Eds. Carol J. Burns and Andrea Kahn (New York, NY: Routledge, 2005), 143.

36 F. John Odling-Smee, Kevin N. Laland, and Marcus W Feldman, *Niche Construction: The Neglected Process in Evolution* (Princeton, NJ: Princeton University Press, 2003). Cited in Nicholas de Monchaux, "Local Code: Real Estates," *Architectural Design* 80, no. 3 (May 2010): 88–93.

37 Anthony Burke, "Redefining Network Paradigms," in *Network Practices: New Strategies in Architecture and Design*, Eds. Anthony Burke and Therese Tierney (New York: Princeton Architectural Press, 2007), 58.

38 Adam Greenfield, "Against the Smart City," Part 1: *The City Is Here for You to Use*, Do projects. New York City: Do projects, Pamphlet 1302, October 13, 2013.

39 Burke, "Redefining Network Paradigms," 71.

40 Carole Collet, "Foreword," in *Alive: advancements in adaptive architecture*, Eds. Manuel Kretzer and Ludger Hovestadt (Basel: Birkhauser Va, 2014).

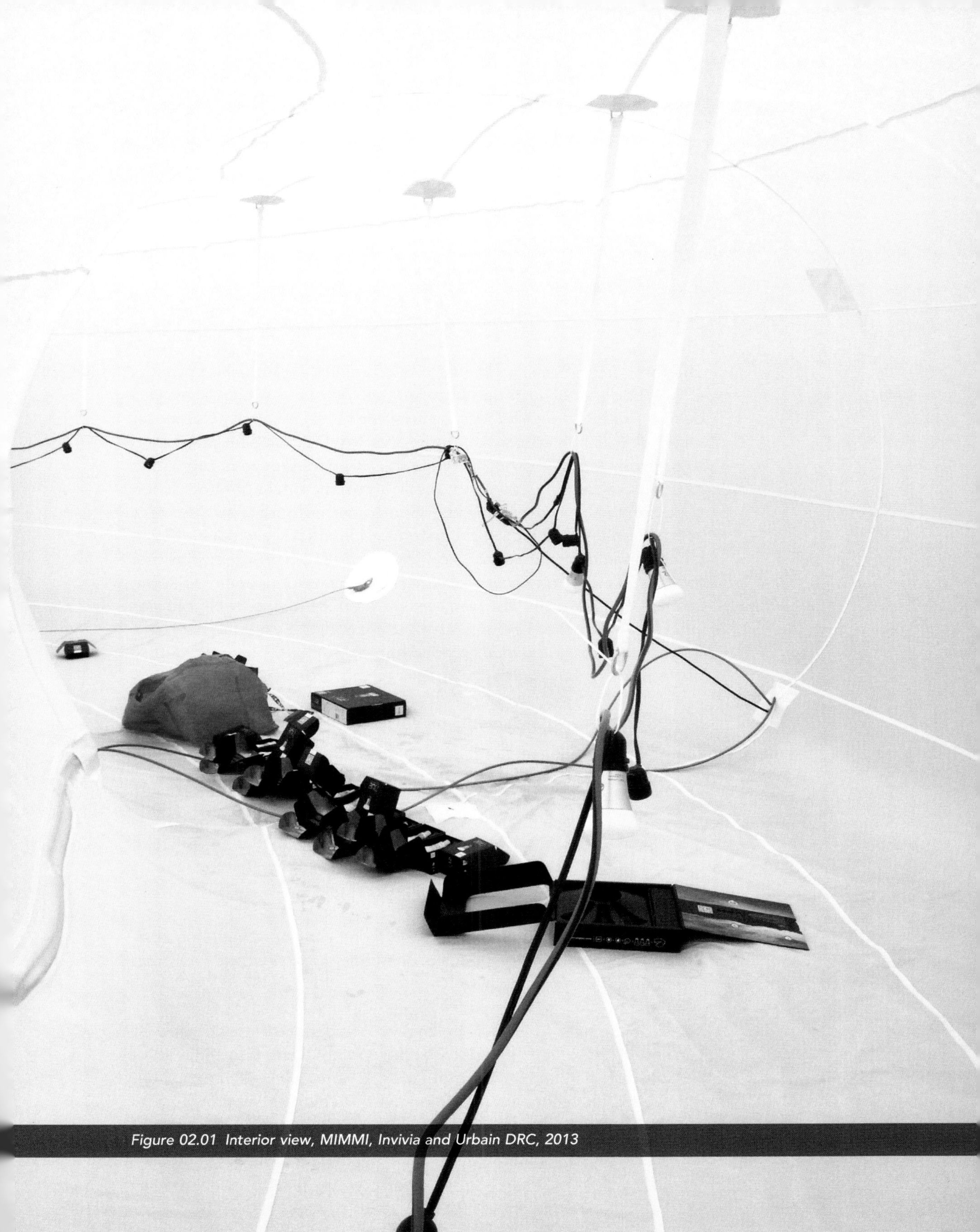

Figure 02.01 Interior view, MIMMI, Invivia and Urbain DRC, 2013

02 RESPONSIVE TECHNOLOGIES

The term "responsive" denotes that an object engages in a process of feedback, a conversation between two actors. In architecture and design this has typically been approached from a Human Computer Interface (HCI) perspective, centering on how human beings respond to or learn from this process. While the HCI perspective is important, it is only a single layer and the view must be expanded when considering the technologies needed to embed responsive systems in environmental or biological systems. This wider view may not even have human beings in it, instead it can be a set of considerations that focuses primarily on ecologies absent of direct human manipulation. Just as importantly, the implementation of these technologies inherently embraces re-purposing and extending by finding new uses for specific technologies or modes of making. Lucy Bullivant, author of *Responsive Environments*, speaks to this broadening:

> . . . the technologies involved, of sensing, computation and display, are in rapid flux, so anachronistic solutions need to be robust; breakdowns are an occupational and institutional hazard, and new schemes are not foolproof . . . designers are extending the versatility of equipment for crafted responsive environments to enable different sensing modalities. The difference is that they customize what exists in order to achieve the right results.[1]

This re-purposing points to new tools formed by hardware and software, generated from the needs of architects and landscape architects to expand their practice.

There are several potentials in landscape architecture supported by current technologies that address environmental and human forms of response. The ubiquity of computation in our daily lives expands the reach of digital technologies, fading them from objects to frameworks that alter our environment physically and perceptually. Methods of sensing and communication support early theories of ubiquity and it is important that a critical stance is developed between local interactions and territorial systems. Anne Galloway speaks to this condition in her article, "Intimations of Everyday Life":

> . . . ubiquitous computing seeks to embed computers into our everyday lives in such ways as to render them invisible and allow them to be taken for granted, while social and cultural theories of everyday life have always been interested in rendering the invisible visible and exposing the mundane.[2]

This invisibility stems from computing that is embedded not only physically but also perceptually. Proposing interaction beyond data will provide methods of engagement for connecting interaction to materials as well as influencing the materialization of systems, thus rendering invisible processes. This is a key component in HCI as it attempts to provide natural extensions of human activities, extending capabilities and responding in predictable and understandable ways. As ubiquitous computing is utilized to interact with ecological systems, there is a similar series of concerns. This begins with how sensing, processing, visualization, and actuation are choreographed.

Sensing as a larger concept refers to the input of data generated from recording or translating phenomena. While a sensing instrument may be universal, the implementation is unique to each site: different deployments will yield varying depictions or narratives. The implementation is a product of the instruments or organs that act as the inputs for information. It is the slight differences in the composition or tuning of these instruments that produces a specific dataset that can be accessed or imaged. This full array of sensing is important as it opens a being or device to a phenomenological connection with the environment. Sensing is often separated from processing and actuation as it produces a generalized dataset that can be discerned from other aspects and may be a product in itself. This sensed dataset is then accessible to be processed and visualized for a multitude of uses.

There is a vast difference between complex sensing networks and designers working with definable inputs—for example, a photocell connected to an Arduino microcontroller. Many of the experiments the design professions are seeing today involve this direct and localized form of sensing, a discrete connection between sensing mechanisms and the data they gather. This paradigm is changing drastically as huge sensing networks are compiling vast amounts of data across many different types of devices. Networks of this type are typically singular in nature and deployed for very specific tasks, such as the gathering of weather data, seismic data, or other forms of global or territorial data. In this manner the data is specific to the logistic or scientific endeavor at hand and is collected specifically for this purpose. As the cost of sensing hardware is getting cheaper these networks and the data they collect is growing exponentially and in many cases becoming less specific. This lack of specificity creates a data space that has become a virtual repository for streams of information, waiting to be accessed. The scale of these monitoring networks and databases constructs a space of "control, not freedom, . . ., and while we enjoy unprecedented access to information and personal communications devices, we are simultaneously smothered by the cloying ubiquity of networks that have no outside."[3] This internalization is a product of the sensing, but more importantly is the political, legal, and cultural paradigms in place that sequester information.

While sensing is gathering information, the ability to process this information emphasizes ways in which the information may be re-purposed, virtualized, or transformed. The processing of information is vital in the operation of responsive technologies as it takes the raw sensed data and builds relationships out of it. This may be as simple as remapping one stream of data values to create a relationship with another stream, or completely transforming the information. The augmentation of information to produce new relationships and realities through hacking is described by McKenzie Wark, author of *A Hacker Manifesto*:

> Abstraction may be discovered or produced, may be material or immaterial, but abstraction is what every hack produces and affirms. To abstract is to construct a plan upon which otherwise different and unrelated matters may be brought into many possible relations. To abstract is to express the virtuality of nature, to make known some instance of its possibilities, to actualize a relation out of infinite relationality, to manifest the manifold.[4]

Processing of information also works with a notion of temporality that builds a relationship between streams of biotic or abiotic information. This is most notable in HCI as the user is closely tuned to the behavior of the system due to an expectation of interactive performance. This ability to sync with physical events, or virtualized events, is often referred to as operating in real-time. The ability for a video game to display 50 frames per second of video or for a door to open when a motion sensor is tripped would be described as operating in real-time. Processing can also happen in a delayed state or information can be pre-processed, and then retrieved for use at arbitrary times. This temporal dimension is incredibly important in the construction of socio-cultural and biotic or abiotic relationships.

The ability to act upon sensed data through processing is related to methods of visualization. How do we see this information as static moments and sequences? How do we translate them into abstractions and complex relationships? Both the collection and the visualization of data are implicit in how we respond to it as it frames our understanding of the information, curating our understanding and response. Because we intend to operate within complex ecological systems, visualization performs as a mediator to decipher a system. This requires multiple overlays that unpack not only the system but also the protocols that govern a system's operation.

> As we operate within the terms of this encompassing material and procedural environment governed by protocol, what we might term a protocology, there remains the issue of visualization. Identifying and understanding a landscape in protocological terms is necessary before that knowledge can be turned into an active design agenda.[5]

The forms of visualization can be multiple, but ideally it sets upon a method for active transformation coherently representing temporal and spatial relationships.

Acting upon the sequence of sensing and processing is guided by visualization and is made tangible through methods of actuation. Actuation is the transformation of sensed and processed data into a form of physical or virtual action. This speaks to a form of physicality in a process, an immediacy where the field in which sensing is taking place is being modified or acted upon. Usman Haque observed that "designers often use the word 'interactive' to describe systems that simply react to input," for example, describing a set of web pages connected by hyperlinks as "interactive multimedia."[6] Response or interaction denotes a full cycle where a phenomenon is sensed, the data is processed, and it is then actuated entering into a feedback loop where the product continues to be sensed, processed, and actuated again. This feedback loop is the basis for responsive technologies and is often built to be as reliable and consistent as possible. This reliability in the feedback loop creates systems that are predictable members of a system of control. This type of interaction is limited. We might call it pushing, poking, signaling, transferring, or reacting. Gordon Pask called this "it-referenced" interaction, because the controlling system treats the other like an "it"—the system receiving

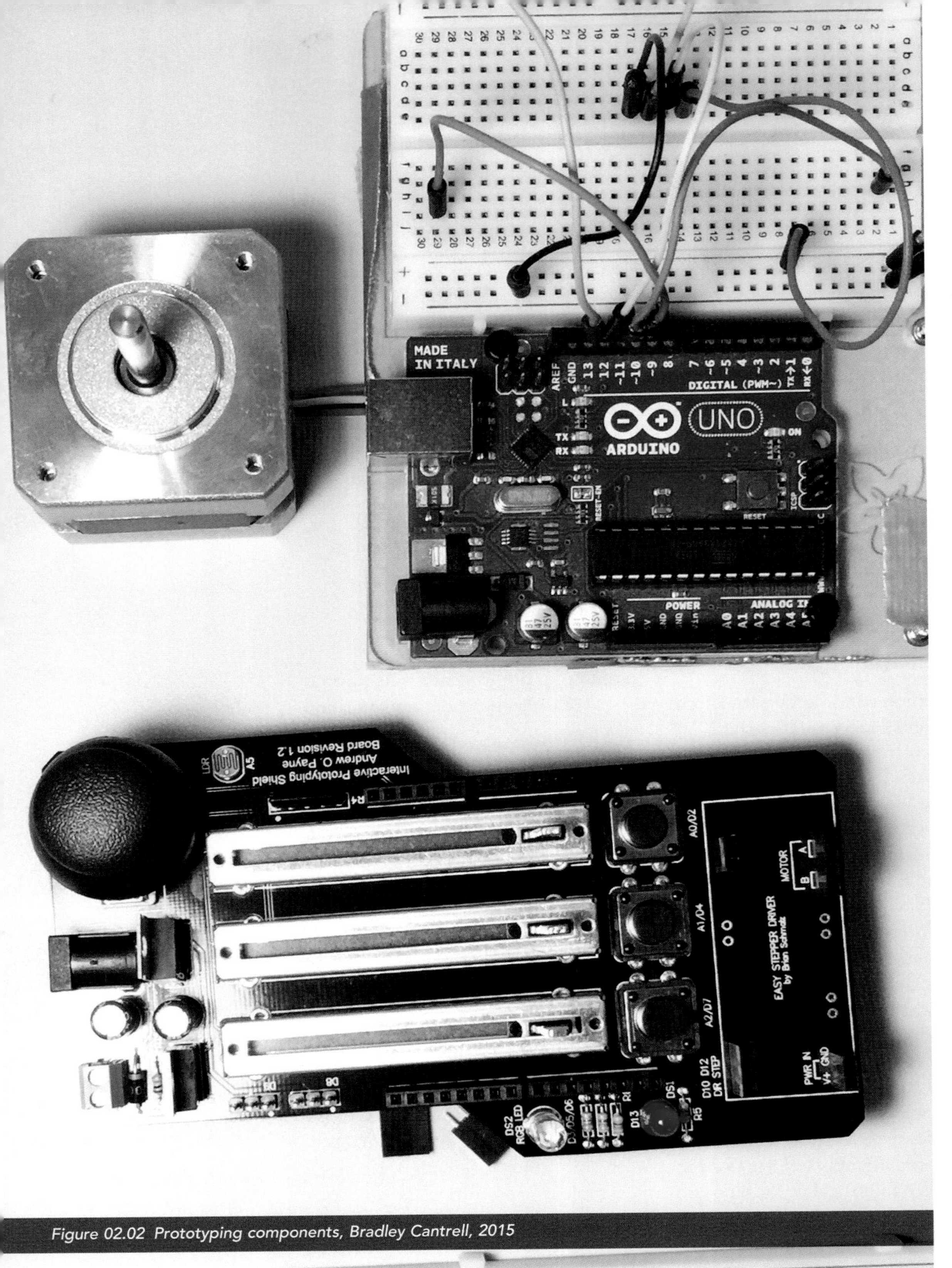

Figure 02.02 Prototyping components, Bradley Cantrell, 2015

the poke cannot prevent the poke in the first place.[7] This inability to provide resistance or counter the process can become problematic in responsive environments and is expanded upon in the discussion on expanding the feedback loop.

TECHNOLOGIES

The action of sensing, processing, and actuation is the product of several technologies that create the ability to develop responsive feedback loops. The first stage, sensing, has a multitude of technologies that can be deployed to create methods to input information. Conceptually, sensing is a switch that detects its current position. While new technologies continually create nascent methods of sensing, often it is the clever re-purposing of existing systems that leads to new sensory modes. A quick overview of sensing technologies highlights sensors for acceleration, acoustics, flow, viscosity, density, motion, optical radiation, orientation, pressure, temperature, electromagnetics, and chemical proportions. It is important to note that, within these categories, the measurement may not be the phenomenon itself but instead an interpretation of the phenomenon to measure another property. There are overlaps in the sensing technologies that allow multiple phenomena to be sensed, depending on the method of deployment.

Actuation is enabled through technologies that alter the physical environment, manifesting itself through transformations. Technologies such as motors, servos, shape memory alloys, and many more provide ways for designers to transform the physical world. These technologies are used to render, regulate, control, and automate environments. Recently within landscape management practices, sensing technologies are used to monitor soil humidity. They are processed to actuate irrigation systems, creating efficient and timely applications of water. This feedback allows homeowners to have gardens, lawns, and landscapes that would otherwise be impossible in certain climates. Putting water consumption issues aside, this is a fundamental change in the way certain landscapes can live within non-native climates or highly disturbed environments. In a similar manner, this method of landscape intelligence is extrapolated through large-scale systems of irrigation that rely on the conveyance of water to grow homogenous crops—managed through autonomous systems this feedback loop is a simple relationship between a single biological need and a constructed prosthetic that supports this need.

PROTOTYPING

In the past decade, the availability of technological tools, access to software development, and hardware prototyping marks one of the largest shifts to increase the ability for designers to prototype and experiment with responsive technologies. This can be seen in micro-controllers such as the Arduino or Raspberry Pi and the integrated development environments that accompany them. Not only are these development environments increasing in accessibility, they are also becoming directly integrated into common modeling and drafting tools with plugins such as Firefly for Grasshopper and Rhino.[8] This direct connection creates links between sensing and actuation with parametric modeling tools, going so far as to remove the necessity for coding and replacing it with a visual scripting paradigm. This collapse in the barrier to entry puts architects and landscape architects directly in control of the prototyping process where they can begin to develop proofs of concept. Beyond accessibility to tools is the proliferation of projects through open-source licensing, which allows each successive designer to build upon previous work. Deconstructing another designer's project to understand their code, hardware solutions, and overall methodology is invaluable and is an ever increasing source of common knowledge.

This ability to develop modes of interaction and deconstruct previous work creates an atmosphere where design through hacking and prototyping thrives.

> Hackers create the possibility of new things entering the world. Not always great things, or even good things, but new things. In art, in science, in philosophy and culture, in any production of knowledge where data can be gathered, where information can be extracted from it, and where in that information new possibilities for the world produced, there are hackers hacking the new out of the old. While we create these new worlds, we do not possess them.[9]

Conceptually this ethos is important and promises a robust methodology of testing and failure that is important to the progress of responsive technologies both in architecture and landscape architecture.

In this spirit, *Responsive Landscapes* presents case studies intended to be viewed as prototypes, tests, experiments, and "hacks." Experiments that lead us further into a deeper discourse on the relationship between sensing, processing, and actuation and how these developing methodologies are transforming our perception of the environment.

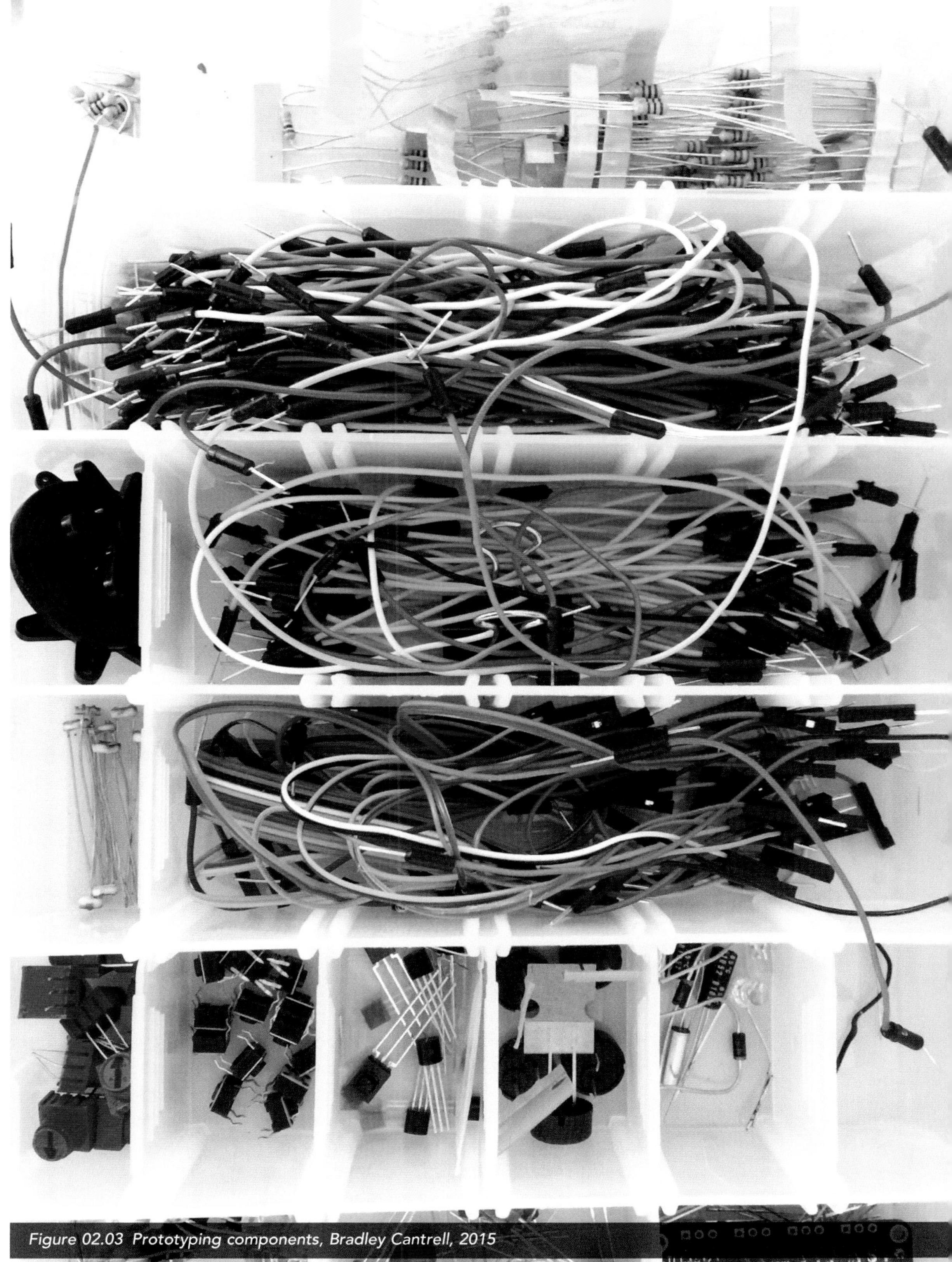

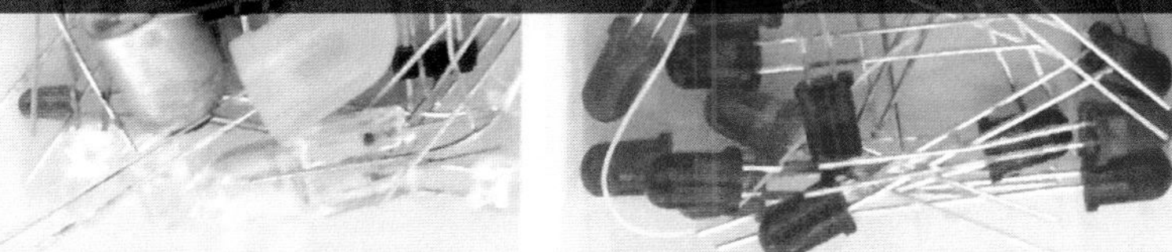

Figure 02.03 Prototyping components, Bradley Cantrell, 2015

> The promise of our evolving supernatural facilities – thanks to a myriad imaginative prosthetic applications of digital technologies – demands that creative practitioners fully involve people in their development on both subjective and objective levels, enabling them to make their own connections between what are increasingly permeable cultural thresholds of perception and being. [10]

The case studies are at once about the technologies used to create them while also firmly framing our evolving view of a changing technological landscape—quickly emerging both inside and outside fields of environmental design. A space that begs to be richer, more diverse, and just through the design of not only culturally and socially significant landscapes but also through their relationships to the world as a whole.

VISUALIZATION, MAPPING, AND SIMULATION

In addition, the accessibility to technological tools creates a new lens that can be used to understand and interact with complex systems. Modeling software enables sophisticated simulation of site phenomena, providing tools for decision making within complex landscapes. The simulation of dynamic systems within the landscape enables designers to visualize and represent data with an increased knowledge of relationships. These models are effective as an interpretable representation because they establish a metric that translates numerical data from simulations into hybrid or coupled models. This metric becomes the underlying fabric of a representation that appeals to our ability to observe pattern and order in both short and long term cycles.

The development of these models creates custom connections between environmental phenomena and modeling or mapping methodologies. These connections are the key for developing common onto-cartographic methods, formats, and processes, which map the agency of things or, as Levi Bryant describes them, "machines." An onto-cartography defined by Levi Bryant is, "in one of its significations, a mapping of relations between these machines so as to discern their lines of force."[11] Machines within this context can be understood as physical materials within the landscape engaged in ecological systems, infrastructures, sensing or measurement devices, and as artificial intelligence, all acting and playing vital roles in shaping contemporary ecologies. Machines by definition are performative: they operate, act, and apply force, they have agency. The onto-cartography or map becomes the simulation and design tool for strategic intervention, an approach that recognizes the agency of technological machines and their connections to evolved biological, political, and economic machines. As ground-breaking as these methods are for the discipline, many of the models or simulations produced are a vast distillation of site systems, lacking the complexities existing within a site. There is a significant disparity between the accuracy of the simulation of particles in design software and many of the more sophisticated flow simulation tools. Operating at this low fidelity requires conscious methods of abstraction for highly complex systems.

When simulating the physical world, we rely on extracted data to develop a working model. The model is understood as a quasi-objective rendering of reality where the data itself is objective and curated for the purpose of creating a model and tools of measurement. These tools or measurement devices are designed with defined goals, associated not only with what it is measuring but also why it is measuring, and therefore creating a "synoptic" view of reality. This abstraction of reality requires a purposeful logic to assign material actors to quantities, values, forces, and locations, among other properties. The model or simulation produced from this logic is then tuned through observations of the physical world and used to manipulate, measure, and quantify selective aspects inevitably causing reality to become a product of the logic, a "selective reality."[12] The history of this condition is identified through the early forestry management practices and agriculture. To establish a metric in early forestry techniques through a "fiscal lens" concerned with overall production, the planting of trees were ordered within a Cartesian grid, simplifying methods for quantification and data extraction. This metric provided data associated with the economic viability of production, but subsequently ignored elements, such as biodiversity, contributing to the resilience of a forest ecosystem. The metric, ordered to easily observe stasis versus change, lead to an oversimplification of complex ecological processes by bracketing certain factors.

To navigate issues with the fidelity of the simulation and create a relational model of ecological dynamics, requires a hybridization of multiple models, pulling them together to function side by side as a composite simulation. There are currently working versions, of simulations bolstered by multiple models such as Google Earth, fluid dynamics models, and climate models. The virtualization of something like climate demands a considerable shift in the way that we are simulating material phenomena and building virtual worlds. A view of climate in the wake of climate change is much more totalizing, it encompasses not just hydrology, but hydrological systems, weather systems, and anthropogenic influences that act on climate. The way each element relates to one another represents the coupling of models that can be viewed as a combinatory projection, and weight the defining inputs to problem-solve through multiple lenses.

This form of simulation is projective, allowing for speculation and inquiry by opening up negotiation between accuracy, projected futures, and intention.

In many ways we are already implementing a slew of responsive systems through modes of production and automation for efficiency. "The architectural profession remains relatively steadfast in a distinction that divides designers from users, even though technology increasingly provides grounds for diminishing that distinction." [13] This can be seen in several cases including climate control, agriculture, and logistics among many others. While the thermostat is an extremely simple device, it creates an important feedback loop that senses then processes through a simple "if: then" scenario, and then actuates. As an architectural mechanism, climate control creates an extremely important feedback loop to maintain an atmospheric equilibrium. Not only does this create space that is comfortable and climate that is convenient, but it also enables whole new modes of construction and program. This form of control opens the possibility for reliable cooling of food or the preservation of archival documents, cases that depend wholly on the feedback provided by sensors and the control of heating, cooling, and humidifying systems. This has evolved and current technologies are able to use predictive modeling to observe usage patterns and create efficient scenarios that attempt to not only optimize for efficiency but also for comfort.

Under a similar paradigm agriculture has evolved tremendously in the past century through automation, precision, and feedback. This evolution has occurred on multiple fronts through new technologies for harvesting and planting, more precise abilities for selective breeding and genetic manipulation, and political and economic efficiencies and exploitations. Responsive technologies are directly involved in each of these areas. Through precision agriculture, responsive technologies are deployed to more accurately plant crops, to take advantage of planting shifts from season to season or to optimize crop layout. Similarly, through the analysis of crop yields and terrain, hydrology, and soil characteristics, a phytogeomorphological approach can be developed to ascertain planting patterns.[14] These are physio-spatial manifestations of responsive technologies, but this can also occur through new synthetic forms of planting media, controlled monitoring, and sampling methods. As these methods continue to evolve they are slowly becoming more nuanced, from sampling per hectare to monitoring single plants. This evolution of precision changes the scale of interaction from the commercial agricultural scale of thousands of acres to the individual plant, making these technologies applicable in urban systems at smaller scales.

Logistics is also a case where responsive technologies are employed to efficiently operate a complex network of systems that adhere to simple, connected goals. On a larger scale than agriculture, logistics has fundamentally changed the way systems are controlled, tracked, and ultimately co-ordinated. This has altered our systems of retail and shipping, creating highly responsive networks that connect disparate elements. The largest retail companies such as Amazon and Walmart are essentially logistics mechanisms, creating extreme efficiencies in distribution. This has developed global scales of operation that are vastly larger than anything seen in human history.

Our environment is saturated with "smart" devices that we use every day. Many of these devices are embedded into our lives in such a manner that we hardly know they exist, functioning in the periphery and blending into the context of our environment.[15] As designers of the built environment, our charge requires us to engage this layer of technology, which is increasingly affecting the environment.

NOTES

1 Lucy Bullivant, *Responsive Environments: Architecture, Art and Design*, 14 (see chap. 1, n. 20)

2 Anne Galloway, "Intimations of Everyday Life: Ubiquitous computing and the city," *Cultural Studies* 18, no. 2/3 (March/May 2004): 384.

3 Alexander Galloway, *Protocol: How control exists after decentralization* (Cambridge, MA: MIT Press, 2004), 147.

4 Mckenzie Wark, "Abstraction/class," in *The New Media Theory Reader*, Eds. Robert Hassan and Julian Thomas (New York: Open University Press, 2006), 213. Previously published as "Abstraction/clasee," in *A Hacker Manifesto* (Cambridge, MA: Harvard University Press, 2004).

5 Burke, "Redefining Network Paradigms," 71 (see chap. 1, n. 37).

6 Dubberly, Haque, and Pangaro, "ON MODELING: What is interaction?: are there different types?" 69–75 (see chap. 1, n. 16).

7 Ibid.

8 Rhinoceros 3d is a 3d modeling application, Grasshopper is a plugin for Rhinoceros that provides a visual programming interface, Firefly is a plugin for Grasshopper that provides direct connections to microcontrollers and external peripherals such as webcams or the Microsoft Kinect.

9 McKenzie Wark, "Abstraction/class," 212.

10 Lucy Bullivant, "Alice in Technoland," *Architectural Design* 77, no. 4 (2007): 13.

11 Levi R. Bryant, "*Onto-Cartography* Author Q&A," *Speculative Realism Series*, Ed. Graham Harman, 2014. http://euppublishing.com/userimages/ContentEditor/1396275575603/Onto-Cartography—Author Q&A.pdf.

12 James C. Scott, *Seeing Like a State: How Certain Schemes to Improve the Human Condition Have Failed* (New Haven, CT: Yale University Press, 1998).

13 Matthew Fuller and Usman Haque, "Urban Versioning System 1.0," in *Situated Technologies Pamphlets 2*, Eds. Omar Khan, Trebor Scholz, and Mark Shepard (New York: The Architectural League of New York, Spring 2008), 13.

14 John A. Howard and Colin W. Mitchell, *Phytogeomorphology* (New York: John Wiley & Sons Inc, 1985).

15 Malcolm McCullough, *Digital Ground: Architecture, Pervasive Computing, and Environmental Knowing* (Cambridge, MA: MIT Press, 2004).

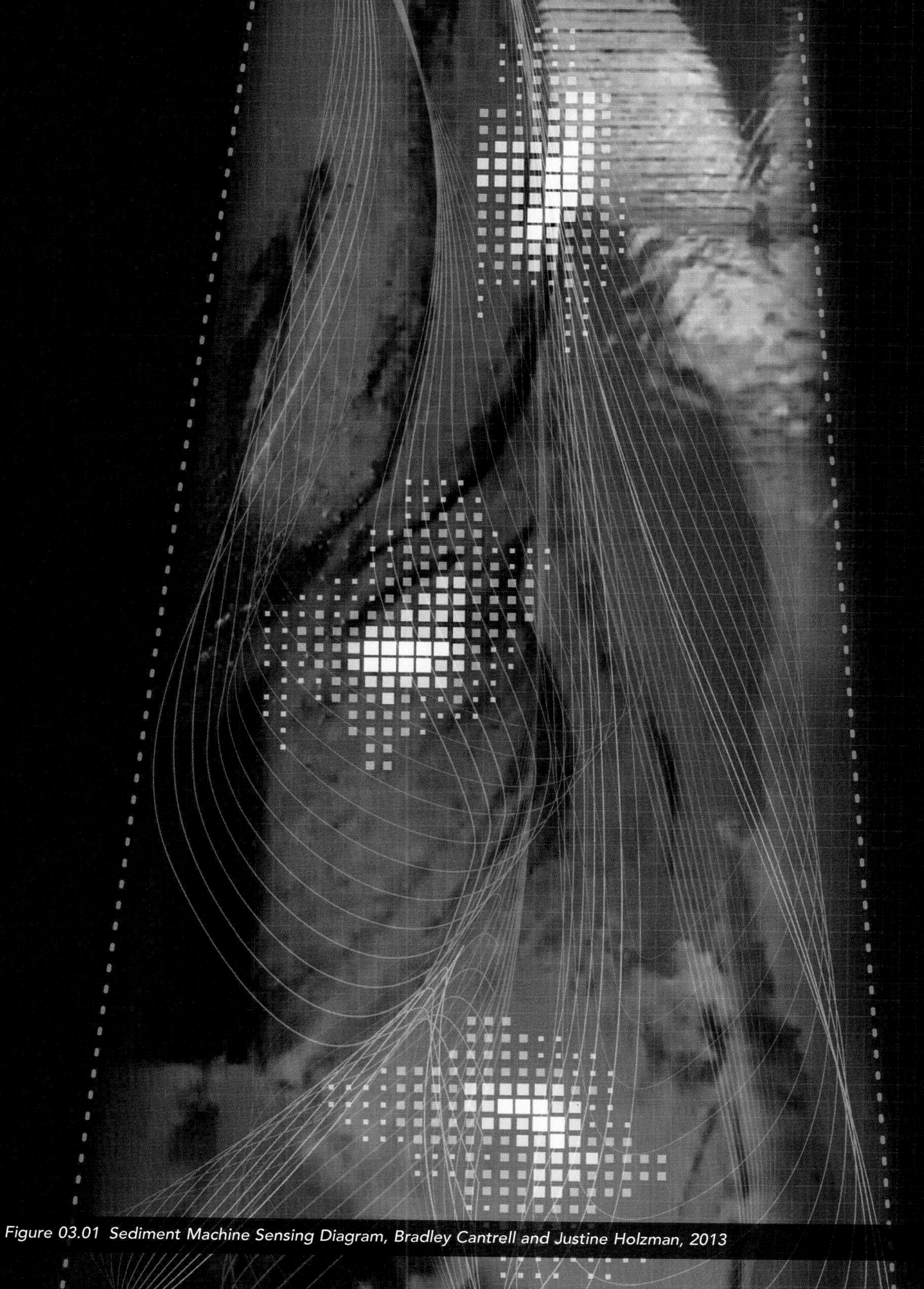

Figure 03.01 Sediment Machine Sensing Diagram, Bradley Cantrell and Justine Holzman, 2013

EXPANDING THE FEEDBACK LOOP

03

Theories of communication between biological and mechanical systems emerged as soon as the foundations of technology appeared. Systems theory was followed by cybernetics, aiming to understand the functions and processes of systems that participate in circular and causal chains—moving from action to sensing to comparison back to action—especially between artificial and biological systems.[1] The investigation between biological and mechanical systems intersects a variety of disciplines, including psychology, ecology, linguistics, anthropology, and visual arts. Similar to the tuning of physical models and simulations, synthetic learning is tuned through historical data and ongoing simulations to define approaches to complex problems. This approach requires two distinct components, clear processes of abstraction and generalization as well as robust interfaces with physical systems. The translation of complex interactions into abstract or generalized models requires an evolving process, where the model is continually evaluated to understand the relationship between fidelity, accuracy, and chance. This interface stems from an abstracted relationship, but must maintain a flexible evolution allowing for new logics to harness the connections that are created. In this manner a relationship develops between programmed logic and physical interaction.

Within the case studies outlined in this text, and much of the work in the field of responsive architecture, technology takes the form of an object communicating with external interactions. The emphasis on the object as mediator, instead of context or environment, reduces the complexity of systems—filtering out fluid connections, varying conditions, and processes that can emerge in a higher order, multiple-loop, or evolving dialogue. Stan Allen formerly advocated this position suggesting that, as a discipline, Architecture should shift focus from objects to "field conditions" to address the form *between* things instead of the form *of* things.[2] It is this liminal zone of landscape (the field) punctuated by architecture that is synthesized by landscape architects. As we design the built environment, addressing the need to control, modify, and interpret ecologies in an effort to enable sustainability and adaptability it is necessary to exploit the intricacies of context as a means to create connections between ecological, cultural, and social systems.

The application of responsive technologies in landscape architecture requires a formal shift in the relationship between user, environment, and computation to focus on context and interrelations. The primary research in computation and interaction, particularly within physical computing have focused on Human Computer Interaction (HCI).[3] The relationship between the user and the machine is an intricate web that alters human perception and mediates between modes of simulation (in software) and forms of articulation (in hardware). This relies heavily on positing that the role of computation centers around human relationships with systems that are both simple and complex. In considering HCI as a device for shaping and mediating physical environments, this anthropogenic approach must be expanded and linked to an interaction model that is contextual. Response to site context is fundamental to landscape architecture, creating metaphorical connections or performative relationships. This form of contextual response becomes embedded within the behavior of responsive systems, creating site specificity through behaviors or outputs rather than solely through metaphor, morphology, or aesthetics.

Feedback denotes a call and response. It is the ability of a system to act, process the actions, and then respond with an updated action. This causal chain of response is the feedback loop and is central to self-regulating or evolving systems. This was discussed briefly in the previous chapter but deserves further exploration to examine the particulars of how this might evolve.

EVOLVING RESPONSE

The feedback loop as a framework denotes an interaction between user and system or device, an updating and cyclical series of events. The common example has been the thermostat, called out early on by the cybernetician Gordon Pask,[4] and described in Usman Haque's Hardspace, Softspace ". . . regulating temperature according to our requirements."[5] Even this example has advanced over time and many current thermostats have evolved to not only control temperature but also to visualize energy consumption and to predict ideal temperature settings. "In the feedback-loop model of interaction, a person is closely coupled with a dynamic system. The nature of the system is unspecified." If this is the case then ". . . the feedback-loop model of interaction raises three questions: What is the nature of the dynamic system? What is the nature of the human? Do different types of dynamic systems enable different types of interaction?"[6] Similarly, from the perspective of landscape architecture, these questions must consider: What is the nature of the non-HCI centric feedback loop? How does feedback change the nature of nature? How do responsive systems function as integral regulators of ecological systems?

As we examine the potentials of these technologies in the environment each of the previous questions is important. Responsive systems that are non-HCI specific ideally expand to incorporate a broader range of constituencies including flora, fauna, and other actors. Broadening the interaction model also broadens the goals that the system attempts to achieve and regulate. The new "natures" that spawn from this approach can be vast, not only evolving in new ways but also performing beyond typical capacities. This form of nature is a product of human designed computational intelligence that evolves in tandem with biological intelligence. These new relationships are inherently unknown but become coupled with the environment. This form of feedback will have the ability to grow in processing capacity, creating a resistant response to ecological systems. It is this resistance that is the primary method of feedback with the potential to create an evolutionary growth between machine and biology. A form of feed-back and resistance is central to the design of the environment; it can be seen in a range of contemporary landscape architecture projects. The planting of 25,000 birch trees at Schiphol Airport in Amsterdam by West 8 creates a tension between infrastructural supremacy and the landscape as liminal connector.[7]

Before going deeper into feedback and response in the environment it is important to discuss some underlying concepts regarding inter-action and feedback. Responsive technologies often rely on self-regulating systems that operate on singular goals or outcomes. The goal defines a relationship between the system and its environment, which the system seeks to attain and maintain. A simple self-regulating system, one with only a single feedback loop, cannot adjust its own goal; its goal can be adjusted only by something outside the system. Such single-loop systems are called "first order."[8] Many typical forms of feedback engage this first order and trend towards regulation based on a pre-determined goal. In computer programming, feedback refers to basic logics: the testing of a condition and the execution of code in response. Testing statements such as if: then and if: else create responses as the code parses information. This fundamental binary language is the underlying root of responsive technologies and is inherent to the creation of software. The hurdle for designers in software comes from developing greater nuance and ranges of feed-back. To develop computational logics that evolve, learn, and expand with biological systems, there is a requirement that these forms of binary logic advance to understand the role of the device and the language that connects the device to its context.

Describing the environment as a series of actors, devices, or machines connotes a formal schema that translates the seemingly inanimate into active objects. The term machine, as defined by Levi Bryant, packages physical forms, political philosophies, and biological processes—among all other things—as agents that possess their own properties and behave through a series of internal and external flows.[9] This is an object-oriented model, where the expression of the machine takes on the characteristics of its relationships. Responsive technologies have this form of agency, creating an ontological relationship through sensing, feedback, and response. This form of design hinges on fields of influence and contextual awareness, described as "intelligent ambience"—an attempt to move beyond the capabilities of the machine and to instead focus on connections and relationships. Lucy Bullivant describes this as a way to emphasize "intelligence and its distribution through an environment" highlighting an "understanding of people's character, behavior, and context."[10]

Understanding the communication between systems as a form of language, either co-ordinated or translated, is a useful conceptual envelope to discuss the transfer of data between designed and evolved objects. This metaphorical language exists in a liminal realm between mechanical or digital devices, and biological or ecological systems. Language, and a process of abstraction, addresses simultaneous modes of complexity through common communication protocols. This comes in multiple forms, from a library for design typologies to a discrete interface between entities. The process of abstraction becomes extremely important, taking infinitely complex organisms or systems and describing them through abstract representations that elicit their fundamental performance.

Data is collected, filtered, reduced, and abstracted in order to become manageable. When data is diminished in this way, valuable pieces of context can be lost and connections are severed. The process of abstraction translates essential elements as a way to analyze or classify the properties while maintaining connections to higher fidelities. By separating characteristics, or hiding details of one or more properties, disparate systems can be arranged and compared, allowing one to concentrate on different concepts in isolation from others.[11] In computer architecture, there is a series of abstraction layers that operate between the physical hardware and the software that runs the hardware. These layers hide the implementation details and differences between hardware so that they can function within the same system. Libraries translate commands provided by the programs into the specific device commands needed by each individual piece of hardware. While unessential details are hidden, they are not lost and can be retrieved in the event of later necessity. Abstraction makes it possible to change the quantity of information represented, allowing a reformulation in a simpler formalism to become possible while preserving the quantity of information involved.[12]

As a metaphor, this form of translation from ecological systems, biology, and software is an intriguing way to imagine a new form of the abstraction layer. The operating system or integrated development environment can be useful metaphors to imagine the meta-space that would enable this form of connection or transaction,[13] essentially an evolving translator or compiler that interprets between hardware (wetware) and software (virtualizations). This ecological abstraction layer is the liminal space that serves as a linkage, a form of integrated development environment for the creation of new forms of biological engineering and ecological management. Systems that find common modes of abstraction begin to create opportunities for larger interrelationships between disparate ecological, biological, social, and cultural modes. This common layer of abstraction creates an overarching opportunity to develop, analyze, or accentuate patterns of linkages rather than design self-contained devices or architectures. Each element expresses itself through physical data with underlying streams of complexity that stem from local inputs but are linked and modified within larger systems. The larger system has no aim of control but instead trends toward local stabilities, adaptation, and modification across subsets of systems.

There are two types of abstraction in computation: control and data. Control, process or procedural, abstraction separates the way a procedure or action is used from how it is implemented and focuses on actions. The external representation is the presentation or interaction that allows programmers to intuitively communicate with hardware. Without a form of control abstraction the process of communicating directly with hardware lacks flexibility, is not efficient, and lacks the intuition of abstract arithmetic operations. Data abstraction separates the elements of behavior that are not critical to the procedure from those that are, and allows programmers to hide data representation details behind a simple set of operations.[14] This creates a contract on behavior between data and code. Anything that is poorly defined or not defined when implemented can change, thus breaking the system. Both types of abstraction can be used discretely or jointly, depending upon the requirements of the system. It is important to understand these two forms of abstraction and to develop a layer of complexity that is exposed to constructed systems.

This layer sets an evolving framework and limits the interaction by controlling inputs and outputs.

A system that focuses on defined methods of abstraction must provide methods for separating qualities of each component and classifying those that are similar providing for moments of overlap and possibilities for communication. The merging of disparate systems such as ecology, technology, and culture requires abstraction methods that link each system, preserving desirable properties while allowing complex disparities to be hidden. Abstraction must provide a method that expands the boundaries of programmed systems, which are currently built for specific interactions and lack the opportunity to create evolving organic interactions. Creating a system that expresses intricate complexities is typically an issue of maintaining access to modes of high data fidelity through time.

> Abstraction may be discovered or produced, may be material or immaterial, but abstraction is what every hack produces and affirms. To abstract is to construct a plan upon which otherwise different and unrelated matters may be brought into many possible relations. To abstract is to express the virtuality of nature, to make known some instance of its possibilities, to actualize a relation out of infinite relationality, to manifest the manifold.[15]

Abstraction affects the interrelation between systems through a reformulation of the subject. Each mode has unique properties but a similarity must be considered in order to maintain data fidelity and flexibility when abstracted. Each of these aforementioned systems can be categorized as input, processing, or output creating a component of a larger inter/reactive system. Inputs are typically sensing systems gathering data from the environment but can also come from storage or database systems depending on the implementation. Processing systems are analytical or transformative in nature, causing data to take on new forms. Output systems are varied but are typically either physical/mechanical, or sensorial visualization systems. What does it mean to abstract any of these systems? Can we envision a robust methodology of abstraction that intends to maintain accessibility while promoting an underlying complexity? How does the composite of the systems form a responsive framework that allows for free evolution while maintaining stability?

Dialogue and conversation occur when these systems exchange information in a continuous process, acknowledging and/or influencing each other's responses, creating a progression of feedback loops. These loops can be either single (closed) or multiple (open).

Figure 03.02 Ecolibrium, *Kim Nguyen, Devin Boutte, Martin Moser, and Joshua Brooks, Responsive Systems Studio, 2011*

In a single-loop conversation, outputs are determined by filtering, and the system feeds information back into itself. In a multiple-loop conversation, new information from each system influences future dialogue and depends upon cycles of response.[16] In a true multiple-loop scenario, the output is not pre-determined but is predicated on evolving contextual input. The multiple-loop scenario can be extrapolated to an under-specified or evolving system ordering a new set of possibilities. Instruction is embedded in each device, machine, object, or infrastructure creating rich interactions with the environment and adjacent systems. Systems coagulate and disperse looking for contextual stability; the framework is in flux with shifting priorities and goals.

INTELLIGENCE

Landscapes are inherently intelligent, the biologies that comprise landscapes have their own individual behaviors, logics, and reasoning that allow these systems to evolve through connectivity and response. Speaking about intelligence in landscape speaks to resistance, counterpoints, and individual behavior, it is not what we think of when we use terms such as smart. Intelligence promotes free will and sentience where "smartness is intelligence that is cost-efficient, planner-responsible, user-friendly, and unerringly obedient to its programmer's designs."[17] Intelligence comes from methods of learning that are difficult to obtain through computation and typically has very little in common with methods of efficiency and obedience.

While metaphorical, in this context the concept of intelligence in landscape refers to a feedback loop between the virtualization of a system (landscape) and the physical, real-time manipulations of an environment. Virtualization refers to the removal of landscape data through sensing and monitoring and recreating landscape as an artifact of processing or reaction. This is a loop of monitoring, processing, and actuation that require methods that create independent intelligences within the physical landscape. The establishment of an intelligent system begins with monitoring. To understand the relationship between biological intelligence and computation it is important to outline some of the methods that have led to advances in artificial intelligence and machine learning.

Machine learning and artificial intelligence are important computational paradigms that provide practical methods to create systems that have the ability to learn, evolve, and respond to complex systems. The term "artificial intelligence" is a larger concept that has evolved through advancements in search, machine learning, and statistical analysis. Although the term "artificial intelligence" was initially coined in 1956 by John McCarthy when he hosted a conference that explored

Figure 03.03 Iterative Feedback, Bradley Cantrell, 2010

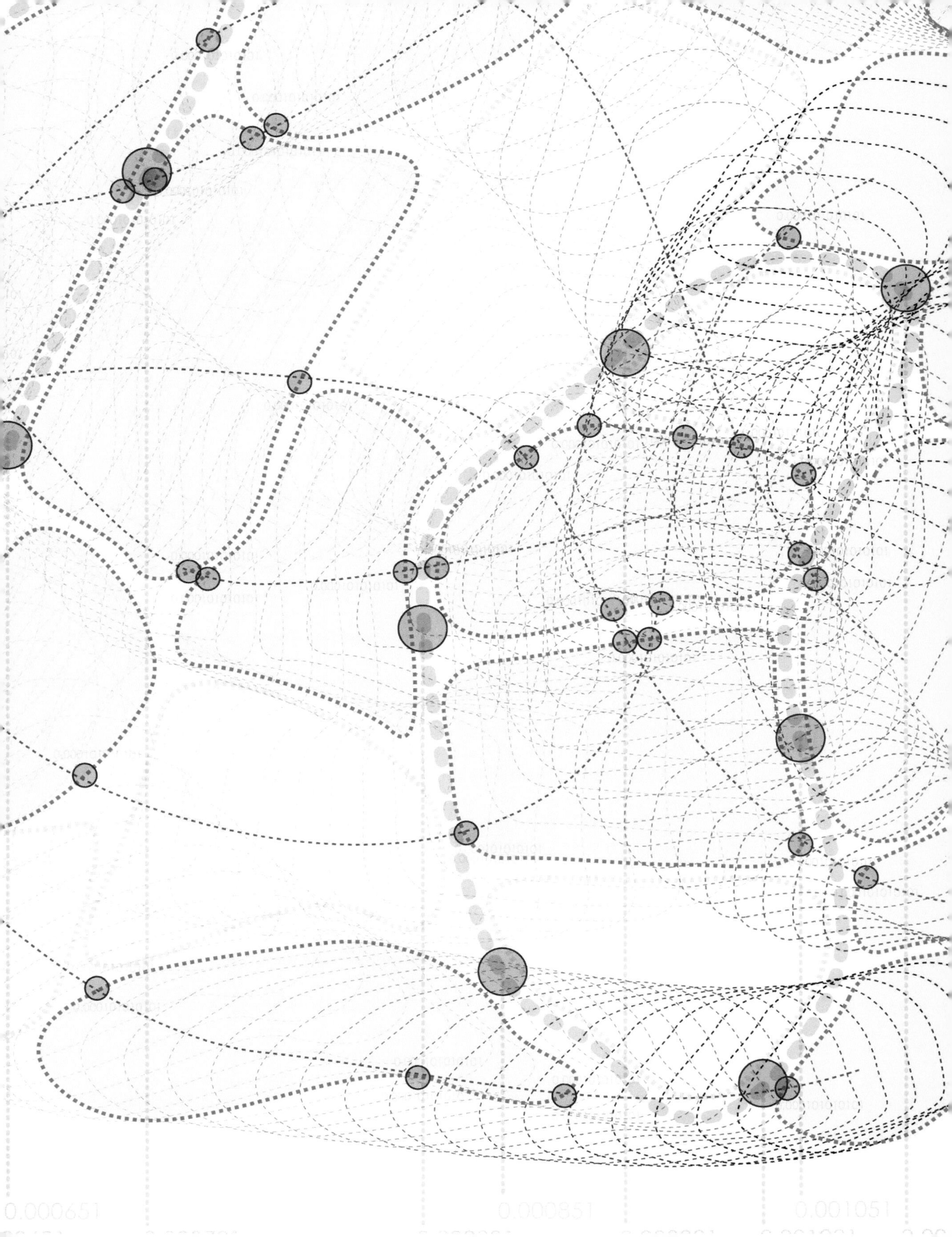
0.000651
0.000851
0.001051

the topic, the idea was proposed by Vannevar Bush's memex, a device that extends human memory and cognition. Alan Turing, in 1950, set forth the question, "Can machines think?" in a paper entitled "Computing Machinery and Intelligence."[18] It is in this paper that Turing introduced the concept of the "Turing test," which was referred to as "the imitation game." The Turing test sets an important, if not pragmatic, benchmark for artificial intelligence that is indistinguishable from human intelligence. Although this form of intelligence has not yet been reached through computation it remains as a continuing goal.[19] Is there a form of the Turing test that is useful for ecological systems? In some ways, this could be said to have been achieved through modeling of simple life forms such as the modeling of C. elegans in the OpenWorm Project.[20] The obsession with the Turing test in the field of artificial intelligence frames human intelligence as the predominant form of sentience rather than imagining new intelligences that may be better suited in a range of contexts.

What constitutes artificial intelligence is continually evolving and the benchmark is always extended. Where many in the 1950s would have declared that a computer that could play chess on par with a human opponent was a form of artificial intelligence, this is no longer the case. Generally, forms of artificial intelligence are subtle, parsing logistics data to determine new efficiencies or giving recommendations for search terms, rather than performing as sentient beings that take on the form and interaction of human proxies. While the estimation of what is to come often outpaces the current state of artificial intelligence, the effect of intelligent systems serves as an extension or mediator of human intelligence. Human memory and processing are offloaded into computational systems creating space for human intelligence to take on other tasks. This intelligence as an extension of humans is a form of feedback that alters our evolution through a form of transhumanism, this becomes apparent as certain technologies become robust and last for multiple generations.[21] A similar form of extension also pushes environmental systems to develop novel ecologies, interactions between digital computation and analog computation. It is this extension of intelligence that is important for ecological systems, it is less important to create computational models that are indistinguishable from ecological systems and more important to extend or respond through new forms of response and computation.

Machine learning is a merger between statistics and artificial intelligence that encompasses a range of methods that allow computers to "learn." In some sense machine learning is a way for computers to automatically write programs rather than being explicitly programmed. This is accomplished through inputs that serve as examples that the computer uses to learn from, these examples then drive the creation of a logic. As a framework, this type of learning creates a way to process a range of data and to create a logic that is dependent upon these inputs, therefore responding and evolving as the dataset or input evolves. One of the first examples of machine learning is from Arthur Samuels who, in 1959, developed software that analyzed the game of checkers. The program Samuels developed looked at which moves tended to win and which tended to lose, and created a strategy based on probabilities of the moves available at each turn. The computer then played tens of thousands of games against itself to learn; this created a form of adaptation and allowed the software to become very good at checkers and to continue evolving based on its opponents. It is this form of adaptation in computation that creates potential for our interface with ecology and environment, an integrated method of analysis and response.[22]

The implementation of machine learning requires the positing of well-formed questions. Forms of learning can be structured such that they focus on experience related to an operation. To evaluate the learning, performance must be measured as it relates to the operation; if this improves or degrades the experience then an inference can be made. The creation of a "well-posed learning problem" was posited by Tom M. Mitchell and Avrim Blum, and varies based on implementation.[23] What becomes readily apparent is that this form of learning can be extrapolated to encompass relationships between infrastructure, urbanism, and ecology. A form of learning that derives new forms relationships that are the product of evolving biotic and abiotic intelligence.

SYNTHESIS

Intelligence is a powerful concept when contemplating landscape systems, particularly when considering feedback and interaction. The project of artificial intelligence gives insight to the larger question of how designed landscapes respond with higher levels of complexity and have a form of agency, evolution, and resistance. The relationship between designed intelligence and evolved intelligence presents an additional tool in landscape architecture's repertoire of media. The complexity of ecological systems exceeds human capacity to understand, manage, and control. As designers and engineers our methods have predominantly approached landscape as a medium that requires control and manipulation, a need to fully articulate the future of place. Another view of this control can frame it as a form of abstraction, simplifying the complexity of the landscape system to provide a method that designers can act upon. This is not a fault of the profession but instead comes from a multitude of cultural and legal traditions that go beyond the scope of this conversation.

As forms of intelligence, simulation, and language are more deeply integrated within landscape a form of sentience develops, an agency that is embedded within objects. A system or infrastructure based on mutual abstraction would be contextually aware through local interactions such as responsive buildings, traffic monitoring, and/or environmental sensors. The system would have pre-determined goals of safety, but evolving goals of management linked across time, location, and content. Creating a contextual narrative relayed through abstraction allows for a multitude of designers to develop robust local systems that are inherently interconnected. Most importantly this interconnected system does not ignore the value of local context but instead exemplifies it and expands its relationships, creating potentials for rich human experience. The objects designed with agency have the ability to imagine infrastructure that is driven by autonomous logics and engages landscape processes through an evolved learning—a constructed landscape that is neither pre-determined nor bounded but requires our engagement. This is a spatial paradigm that disengages with previous computational roles and requires models that are deeply embedded within ecological systems. Models are abstractions, abstractions that grow their relationships to the ecologies they are modeling.

Usman Haque's description of the "collectively designed project" as incorporating "conflicting logics" refers to numerous intentions competing within the constructed environment.[24] In this sense, the feedback loop (for the built environment) should frame unintentional influence over evolved environments. The larger question lies in how feedback evolves to consider these external consequences, the areas that are not pre-determined but are resultant from the actions of the system. As a method this embraces landscape as a composite of intelligences, a biological form of memory and processing that shifts or opens possibilities. In much the same way we are programming and tuning the landscape, other actors—such as sediment or mosquitos—are doing the same thing. We imagine their input as having less intention, although it has every bit as much agency and outcome. In a similar way that intelligence, as described by computation, is an extension of humanity, it is also an extension of ecology. Responsive landscapes point towards a realm where built systems emerge from goals that are not solely concerned with humanity.

Goals are an integral benchmark in design, an outcome that designers and engineers attempt to match or calibrate toward. Without goals there is no known outcome and iteration relies on the moment, there is only history to repeat. At the same moment, goals promote stasis in that they specify an outcome and the system, or landscape, regulates itself to match this outcome. Conceptually the evolving feedback loop refers to methods of interaction that move beyond static goals and system efficiency and embrace divergent or bifurcating futures. The evolving feedback loop leaves room for novel or emergent conditions. In a critique of Landscape Urbanism's "process discourse" to actualize landscapes exhibiting truly novel and emergent conditions, Julian Raxworthy's dissertation "Novelty in the Entropic Landscape" examines methods in gardening to find material-based interaction and real-time responsive behaviors to novel conditions found in the evolving garden landscape. Raxworthy offers the term "tendency" to describe an approach to designing for change or emergence:

> . . . that seeks to aim towards an outcome rather than concretely specify it. Instead tendency promotes flexibility in how an outcome will result, exhibiting novelty in the form of specificity rather than contrast. Feedback describes real-time processes, such as gardening, that allow for a recurrent involvement in the development of projects over time, maximising emergent opportunities. [25]

This concept of feedback points to a need, particular in landscape, to develop fuzzy outcomes that operate within a range—where success is not narrowly defined, but instead connotates novel outcomes that may deviate from a previous hypothesis. A focus on indeterminate systems relies on unorthodox methods of measurement for success, implementation of failure, and modes of resistance. The indeterminacy of site systems and networks constructs a future for networks as political ontologies, which place the material significance of networks as a critical indicator to establishing effective protocol; the ability for political control of networks.[26] This methodology of applying protocol to biological networks, termed "protocological control," to aid in political resistance can be paralleled to the necessary methods for adjusting the complex networks that would exist in a synthetic or responsive landscape. Small adjustments to protocol, termed "counterprotocols," can be envisioned as a set of design rules, management guidelines, or instructions for landscape manipulation to aid in adaptive management. "Resistance" is generally a political term, used to describe tactics for initiating political reform and can be understood as a metaphor for small adjustments to the system. The idea of resistance is especially compelling within the context of designing counterprotocols for ecological systems and networks. Instead of instituting a definite system or network, the counterprotocol takes advantage of an ecosystem's adaptive and generative capabilities to excite change by a catalytic resistance. This target of resistance as a stimulus for protocol's ability to "sculpt" and "inflect" allows for small adjustments to effectively manipulate the overall structure of the network over time.

Conceptually this is an attempt to answer a fundamental question in contemporary landscape architecture theory, "how can we design landscapes that engage dynamic systems and adapt to changing conditions?" Protocological control and modes of resistance are convincing metaphors for manipulating technological networks but they also have significance for ecological networks. As our technologies increasingly interpret landscape as a network or network of networks, the view of ecological networks can be negotiated through an evolved form of feedback. In relation to landscape, "interactive architectural design . . . will enable the relationship between building and program to become a much more subtle and communicative process, embracing a wider, personalised set of functions, desires and experiences."[27] Pulling back from architecture there are several methods of interacting on landscape that look past personalization and desire, instead highlighting ecological fitness and robust dynamics. The methods described in the following chapters—elucidate, compress, displace, connect, ambient, and modify—employ forms of response, actuation, and perception as actors in the landscape. They are ways to conceptualize both virtual and physical transformations of the environment. Landscape shifts are enabled by the manipulation and choreography of landscape phenomena and matter, these are methods that speak to connectivity.

This is not a new conversation for landscape architecture; pragmatically it examines new ways to shape the environment and is not centered on new forms of human experience. This is a form of landscape that conceptualizes a cyborg—an integrated whole that is formed from integrated processes that are biotic and abiotic. The cyborg speaks to a smartness that goes beyond an environment laden with ubiquitous computing devices:

> The cyborg is the new contemporary archetype, altering existing ecologies by overlaying new sets of relationships between organism and environment. Despite the unprecedented proliferation of ubiquitous computing in daily life, "smartness" in architecture eludes the synthetic promise of the cyborg. [28]

This is a landscape embedded with intelligence and agency, human actions weighed within a matrix of competing interests. A landscape that is difficult to discern humanity from some form of other. It is a network of actors and requires designers to engage this system with methods that embrace adaptive and resistant management scenarios. The promise of a responsive landscape aspires to a methodology that is on one side technological and the other integrative.

NOTES

1 Gordon Pask, *Encyclopaedia Britannica* 14th ed. s.v. "cybernetics," 1972.

2 Stan Allen, "From Object to Field," in *Practice: Architecture, Techniques and Presentation* (London/New York: Routledge, 2008), 251. Revised and expanded from Stan Allen, "From Object to Field," *AD: Architecture After Geometry* 127 (1997): 24–31.

3 McCullough, *Digital Ground*, 154 (see chap. 2, n. 15).

4 Gordon Pask, *Conversation Theory: Applications in Education and Epistemology* (Amsterdam and New York: Elsevier Publishing Co., 1976), cited by Usman Haque in "The Architectural Relevance of Gordon Pask," *Architectural Design* 77, no. 4 (2007): 54–61.

5 Usman Haque, "Hardspace, Softspace and the possibilities of open source architecture," 2002, http://haque.co.uk/papers/hardsp-softsp-open-so-arch.PDF.

6 Dubberly, Usman, and Pangaro, "What is interaction?" (see chap. 1, n. 16).

7 Adriaan Geuze and Maarten Buijs, "West 8 Airport Landscape: Schiphol," *Scenario* 04 (Spring 2014), http://scenariojournal.com/article/airport-landscape/.

8 Dubberly, Usman, and Pangaro, "What is interaction?"

9 Levi R. Bryant, *Onto Cartography: An Ontology of Machines and Media*, in *Speculative Realism Series*, Ed. Graham Harman (Edinburgh: University Press, 2014).

10 Bullivant, *Responsive environments*, 1 (see chap. 1, n. 20).

11 Nell B. Dale and Chip Weems, *Programming and problem solving in Java* (Boston: Jones and Bartlett Publishers, Inc., 2007).

12 Fausto Giunchiglia and Toby Walsh, "A Theory of Abstraction," *Artificial Intelligence* 56 (1992): 323–390.

13 Fuller and Haque, "Urban Versioning System 1.0," 19 (see chap. 2, n. 13).

14 Dale and Weems, *Programming and problem solving in Java* (see chap. 3, n. 11).

15 Wark, "Abstraction/class," 213 (see chap. 2, n. 4).

16 Usman Haque, "Architecture, interaction, systems," *Archiquitetura & Urbanismo* 149 (August 2006).

17 Andrew Ross, "The New Smartness," in *Culture on the Brink: Ideologies of Technology*, Eds. Gretchen Bender and Timothy Druckrey (Seattle: Bay Press, 1994), Series: Discussion in Contemporary Culture, no. 9.

18 Alan Turing, "Computing Machinery and Intelligence," *Mind: a Quarterly Review of Psychology and Philosophy* 59, (January 1950): 433.

19 The passing of the Turing Test is a debated topic and several forms of computational intelligence have passed different benchmarks. The most recent has been an artificial intelligence, Eugene Goostman, created by a group of hackers out of St Petersburg, Russia. The Goostman bot successfully passed the Turing Test by convincing 33% of the judges at a competition hosted by the Royal Society at the University of Reading on June 7, 2014. However, this is not seen as having reached the benchmark set.

20 OpenWorm (website), http://openworm.org/.

21 Nick Bostrom, "A History of Transhumanist Thought," *Journal of Evolution and Technology* 14, no. 1 (January 2005). http://nickbostrom.com/papers/history.pdf. Accessed February 21, 2006.

22 Richard S. Sutton and Andrew G. Barto, *Reinforcement Learning: an Introduction* (Cambridge, MA: MIT Press, 1998), 267.

23 Avrim Blum and Tom M. Mitchell, "Combining Labeled and Unlabeled Data with Co-training" (Proceedings of the Eleventh Annual Conference on Computational Learning Theory, COLT 1998), 92–100, doi:10.1145/279943.279962.

24 Usman Haque, "Hardspace, Softspace and the Possibilities of Open Source Architecture," Haque Design Research, 2002, www.haque.co.uk/papers/hardsp-softsp-open-so-arch.PDF January 19, 2015.

25 Julian Raxworthy, "Novelty in the Entropic Landscape: Landscape Architecture, Gardening, and Change" (Doctoral Dissertation, University of Queensland, 2013), 20–21, (espace) http://academia.edu/8879242/Raxworthy_Julian._2013._Novelty_in_the_Entropic_Landscape_Landscape_architecture_gardening_and_change_School_of_Architecture_University_of_Queensland_Brisbane, July 2015.

26 Alexander Galloway and Eugene Thacker, "Protocol, Control, and Networks," *Grey Room* 17 (2004): 6–29.

27 Bullivant, "Introduction," 7 (see chap. 1, n. 1).

28 Yoon and Höweler, *Expanded practice*, 132 (see chap. 1, n. 18).

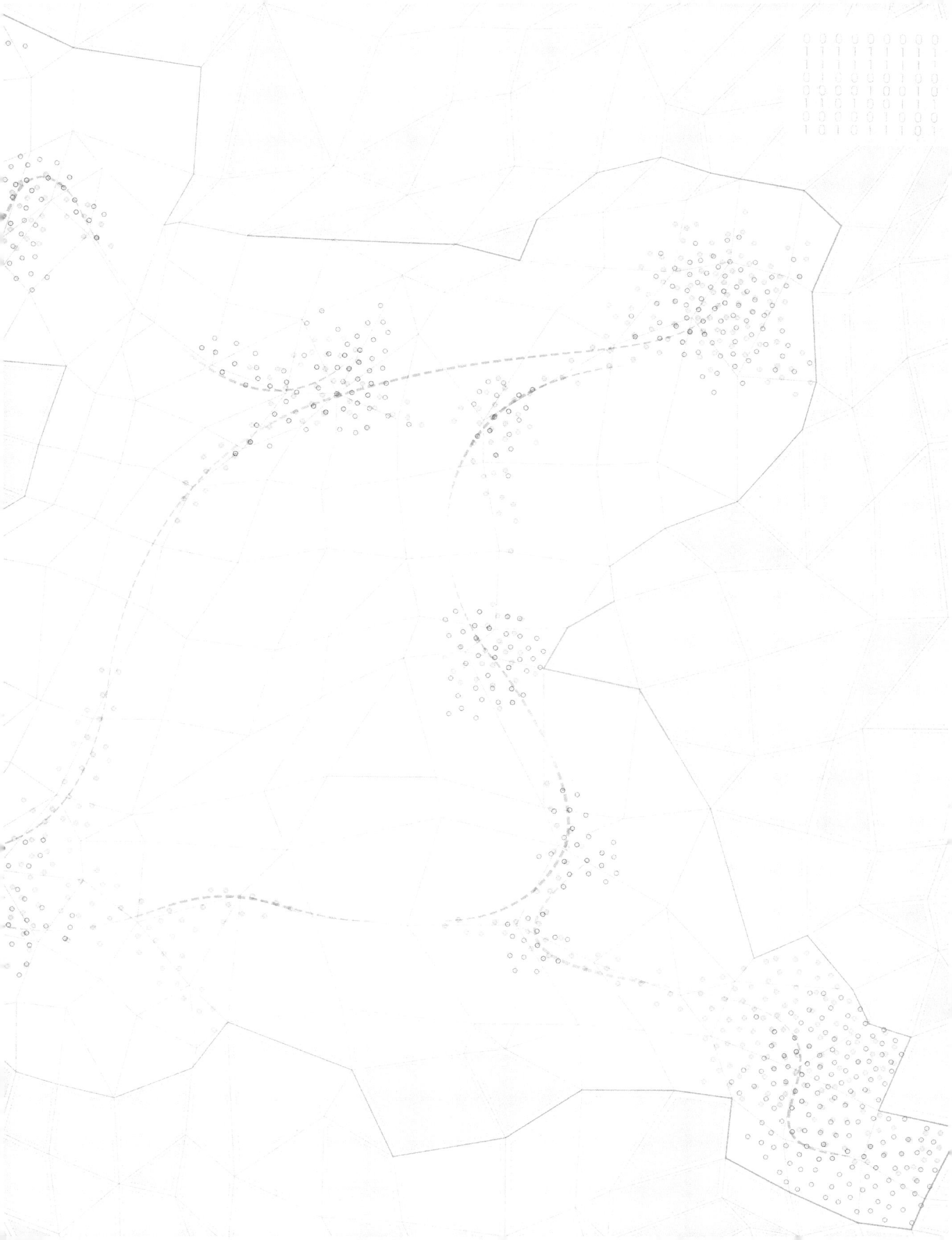

04 ELUCIDATE

A large part of the world is hidden beyond our human sensory capacity. This may be due to our specific range of hearing, seeing, tasting, feeling, or smelling, beyond which lies a world that is perceived through extra sensory mediation. This veil can be lifted or made less opaque by expanding our sensory range, allowing other wavelengths into our visualizations, expressing frequencies beyond our hearing, or illustrating particles that are not felt. This enlightens us to a landscape rich with phenomena, beyond our senses, functioning as a composite to create the world we understand.

Augmenting our sensory palette through exposure to extrasensory projections opens opportunities for interfacing with imperceptible phenomena. Rendering imperceptible phenomena is most effective through an intermittent or continuous relationship, in which the translation of invisible information is learned over time. Responsive projects that *elucidate* bring clarity to ordinarily unseen and invisible phenomena through methods of visualization and actuation. Most projects exhibiting responsive technologies share this component; however, projects within this chapter draw specific attention and intent to interpreting and elucidating imperceptible phenomena into the reach of human senses.

The range of human senses dictates how we experience and operate within the world. Individual senses render the world in different ways, translating many types of phenomena, in which experiential and perceptive qualities of the landscape are formed through a composite of senses. Technological advances in digital photography and film in combination with personal computing have shifted our experience of the world to primarily visual means, transforming the image, "now made of bits," "increasingly malleable," infinitely reproducible, and transferred "like viruses."[1] The proliferation of "these versions and recombinations capitalize on the presumed veracity of the camera to present altered instances as photographic reality."[2] The introduction and advancement of Graphical User Interfaces (GUI) have continued to prioritize the sense of sight. Through personal computers, devices, and sharing, the digital image is infinitely reproducible. Within this digital culture, which is highly dependent on the visual and methods of abstraction, are efforts to bring back environmental phenomena into our experience of digital interfaces. Hiroshi Ishii, founder of the Tangible Media Lab at the MIT Media Lab, has been developing what he calls the "Tangible User Interface" through prototypes that re-integrate multi-sensory and material interactions into digital devices, "the key is giving a physical form, a tangible representation, to information and computation."[3] By coupling mechanisms of control with both physical (tangible) and projected raster imagery (intangible), Ishii hopes to extend human computer interfaces to engage and extend human capabilities.

Within the exploration of how to extend HCIs to encompass multi-sensory interaction, there are broader relationships between human sensory capabilities and external environments to consider. There are obvious limits to human sensory capabilities—what could be considered human blind spots. It is clear that domesticity and urbanity have reduced our sensory capabilities for survival, providing the time and knowledge to develop particular senses over others. However, there have always been limitations to human senses linked to human evolution. For example, humans are only able to hear within a range of frequencies. This limited range is made clear through the extended hearing capabilities and communication patterns of other species. There are sound waves that humans are unable to hear that other mammals are attuned to, dogs hear the early rumblings of earthquakes or high pitched whistles. In the electromagnetic spectrum, shrimp can see ranges of ultraviolet light. There is a wealth of information in the environment that occurs all around us, some of which we are attuned to or have learned to become attuned to, but for the most part are unable to sense or interpret.

To *elucidate* or bring into focus what enters into the periphery or escapes our senses entirely, requires extending the range of human senses. For thousands of years, humans have been developing devices to read beyond our perception and extend our human capabilities without digital technologies. Many of these devices have been discovered or invented to read environmental phenomena in order to track time, explore, find resources, sense danger, or heal illness—all technologies associated with human survival. As these technologies have evolved, their applications have expanded. Some of the most influential technologies were developed in times of crisis, such as war. Immediately after crises have subsided, capital comes into play to further develop nascent technologies into commodities.

Stepping back to the invention of extra-perceptory tools—many sensing devices are developed through known relationships by observing cases of cause and reaction, recognizing relationships between dynamic and material forces. A windsock or weathervane translates wind energy into a visual indicator of speed and direction. Hot water kettles translate temperature into an auditory cue. Kinetic instruments and devices include early inventions that begin to interact with magnetic and electromagnetic fields. Instruments and technological devices used for exploration and travel fundamentally changed how humans viewed the landscape, and have had a profound effect on measuring, mapping, and recording landscape information.

Equally influential are observations of reactions between environmental phenomena and other life forms. Both flora and fauna are potent indicators of external forces ecosystem dynamics, and evidence of contaminants. Many of these relationships are chemical and manifest in changes of color, sickness, death, growth, formation of growth, and contextual cues—such that, biological life itself can be utilized as instruments and devices. A common example would be using certain species as environmental indicators of ecosystem health. In the early days of coal mining, miners would bring canary birds with them into mines as an indicator of high levels of carbon monoxide. The canaries would pass away if there were unsafe levels of carbon monoxide, prompting the miners to exit the mine. Unintentionally, fish kills and algal blooms are indicators of high nutrient content in water bodies.

The function of biological indicators gives way to the design and invention of devices and instruments that perform the task of these biological capabilities. Biomimicry is utilized in sensor designs by intensely researching and reproducing sensory functions of other organisms. The "whisker sensor," modeled after different types of mammal whiskers, is a crucial element for the performance of mobile and autonomous land robots. Current research of robots designed to operate within underwater environments have been interested in pinniped whiskers, found on seals and sea lions, for their ability "to track hydrodynamic trails left by potential prey."[4] Researchers are using biomimicry to emulate the "mechanical transduction of fluid-excited signals by whisker-like sensors and the role of the whisker mechanical structure in the sensing process."[5]

There are also examples of actually extending human perception through biological organisms. Sensing can be extended through biological organisms as they are, in cases like the canary bird. However, there are examples of targeting biological capabilities by designing alterations, either through training, genetic modification, synthetic biology, or prosthetics. Another example, used by Anthony Dunne in his text *Hertzian Tales*, is a "'physiological' receiver that uses the electrical sensitivity of the frog's leg" designed by Lefeuvre as an example of "early meteorological equipment used to make visible atmospheric phenomena otherwise too subtle for our bodies to sense."[6] June Medford has been conducting research within the field of synthetic biology to design plants to detect bombs by changing color in the presence of substances such as trinitrotoluene (TNT), essentially turning the plant into a switch. The change in color becomes easily detectable by the human eye. After designing a sensor and response system uniquely attuned to the plant's biology, the sensor proteins are programmed in a computer, then "encoded in a plant, and empowered with such properties as memory and amplification."[7] The potentials for synthetic biology offer opportunities to "design traits that are new to evolution and beneficial to humanity."[8]

Similarly, Chris Woebken, a collaborator on the project *Amphibious Architecture*[9] has been interested in extending the sensory capabilities of animals, what he terms "animal superpowers." In an interview with Geoff Manaugh featured in *Landscape Futures*, Woebken describes his excitement over current research utilizing animals as biosensors, particularly the work DARPA is doing: "the moth has a superpower to detect pheromones, for example, and you can actually train them . . . they can twist the pheromone communication, using sugar water, so that the moths will look for explosives."[10] Woebken's work seeks to understand the potentials of interacting with biological organisms: "how you can use their superpowers to create sensors that don't just allow you to use them in a military sense, but for other projects."[11]

Some of the designed devices for extra perception are quite ubiquitous, fundamentally altering fields of study. The methods in which we navigate our environments both internally and externally have been forever augmented by technologies such as X-ray imaging, infrared, and sonar. The X-ray machine images the internal human body through electromagnetic frequencies. Anthony Dunne describes the technology as "a sort of radio perspective, revealing, concealing, and exposing hidden organs and views, and creating a 'radio theater' of the hidden body," illustrating the human body as "an electromagnetic medium."[12] The introduction of satellite imagery and geographical positioning systems has profoundly elucidated the landscape. The coupling of these innovations with remote sensing technologies has changed our capacity to see the landscape and analyze landscape conditions. Agricultural practices are being reshaped by the combination of GPS, remote sensing, and precision agriculture technologies to produce incredibly precise mappings of crop yields. Because harvest combines collect a GPS location every square foot, farmers have the information to make nuanced management decisions based on the productivity and performance of their soils. For instance, it may be more profitable not to plant in certain areas because the cost is higher than the yield.

LIDAR, Light Detection and Ranging, has fundamentally altered our ability to track, compute, and run analyses. Data is collected through a pulsed laser that measures ranges of light frequencies to the Earth that, in conjunction with global positioning systems, produce highly accurate and 3-Dimensional (3D) readings of topography and material qualities of the landscape. The open-source products of these technologies have initiated many interesting and perhaps unanticipated and positive utilities associated with previously imperceptible patterns of land use. Infrared has been used to find cases of illegal deforestation previously undetected within the Amazon Rain Forest, through reverse imaging techniques.[13] The Greek government has employed simple searching on Google Maps to find residences with personal pools who have not paid the associated taxes, finding the astonishing number of 16,974, in comparison to the 324 reported pools.[14] The Satellite Sentinel project has been prosecuting war criminals and fighting against genocide in Sudan through an analysis of DigitalGlobe's[15] satellite imagery, to identify "elevated roads for moving heavy armor, lengthened airstrips for landing attack aircraft, build-ups of troops, tanks, and artillery preparing for invasion—and sound the alarm" to report undocumented and unreported crimes with evidence indicating "alleged mass graves, razed villages, and forced displacement."[16]

The utility of these devices to elucidate imperceptible phenomena stems from the combination of methods in measurement, distillation, and translation. The translation of phenomena may be a form of abstraction, computation, or comparison. Alterations of methods elucidate very different representations of unique phenomena. Devices and technologies designed for specific purposes can be appropriated or hacked to elucidate or interact with phenomena in novel ways. Natalie Jeremijenko's project *Feral Robotic Dogs* started in 2005 as a course with Yale Students titled "Mechanical Engineering 386, Feral Robotics: Information Technology in the Wild" in which the students were tasked to hack cheap children's robotic toy dogs—originally designed to playfully walk and bark—and assign them an alternate function.[17] The students altered the devices to sense, or more appropriately "sniff," contaminated soils in urban areas and bark. The toy dogs were networked so that when one device found a toxic area, all the toy dogs were alerted to swarm the area and bark, thus marking the toxic territory and potentially elucidating an area that had been considered "safe" by government standards.

The opportunities for re-appropriation of devices and subsequent data are endless; however, the methods in which these devices are deployed actively shape our understanding and interaction with the landscape.[18] In speculating about landscape futures, Geoff Manaugh was taken by how existing instruments have influenced and shaped our conceptions of landscape: "even lumbering and lonely Mars rovers packed with instruments for off-world exploration—also reconfigure, albeit in very different ways, our existing understanding of a given landscape."[19] Visualizing this new world of unseen or imperceptible phenomena is already happening.

> The world was already filled with extraordinary instruments, handheld sensors, mechanisms, experimental arrays, and other, often semi-autonomous, networked machines through which humans, on a continual basis, without pause, on every continent of the Earth and even at the bottom of the sea, have been recording and interpreting the world around them. [20]

The concept of *elucidate* suggests opportunities for landscape architecture to engage in the spatial design and rendering of extracted and translated information by building from known methods. These methods include using landscape interventions to augment information, build narratives from complexity, and activate processes for engaging and unpacking data. The individual device, networked devices, transmission, radio, satellite, and internet all offer tools for elucidating landscape conditions.

Elucidation is a fundamental form of the operation of responsive technologies, at the core of all of the forthcoming topics in the case studies. With the introduction of radio transmissions, telecommunications, and radar, "The twentieth century has seen space evolve into a complex soup of electromagnetic radiation. The extrasensory parts of the electromagnetic spectrum form more and more of our artifactual environment . . . "[21] Anthony Dunne describes how designers render material conditions through "visual motifs" and "representations" and only on occasion interact with material conditions "directly as a physical phenomenon."[22] The field of Landscape Architecture deals with both the material world and the immaterial world of representations used to understand and manipulate the landscape. *Elucidate* intends to bridge the material and the immaterial, crafting distinct relations between the representational and the actual.

There's a distinction between direct and mediated methods for translating phenomena into an elucidation. Direct devices, like the windsock, function in a way where we are able to understand the relationship of the device to the phenomenon because it is embedded in the function. The windsock itself becomes the sensor. This phenomenon is translated through the device and so, in a sense, the device itself becomes the representation. Mediated devices extract data then represent it using graphic or spatial methods to elucidate phenomena. Here, the data is extracted, processed, then actuated to render the phenomena. In this scenario, the representation of phenomena may be completely detached from the phenomenon itself, communicated only through abstraction. Within these distinctions between physical and biological, direct and mediated, there are endless hybrids and connections to be made in order to elucidate various phenomena through spatial, particularly landscape scale, settings through the mixture of material and mechanical processing.

Shannon Mattern warns of the existence of "raw data," stating that there is always "aesthetic dimension to their derivation and presentation" and laments "the exhausting ubiquity of data visualizations" in which we see through citizen science and amateur design proposals, "the rise of an aesthetics of measurement."[23] The concept of *elucidate* is not meant to proliferate forms of ambiguous data-visualization, rather it's to couple ongoing ecologic processes with designed interventions in a way that illustrates their complexity and evolving relationships, such that meaning may be extracted. However, often what is found beautiful or aesthetically intriguing is a scene of environmental sickness or destruction.[24] Numerous oil paintings of sunsets emerged around the beginning of the industrial revolution. The pollution in the air at that time produced gorgeous sunsets, recognized by landscape painters. Although people were unaware of the consequences of pollution, they took notice of the shifting landscape conditions. Edward Burtynsky's images of strip mines, piles of coal, factories, and disturbed hydrological systems are famously beautiful, while warning of irreversible change. Kristina Hill finds hope in the power of landscape aesthetics to communicate the new landscape sublime of sea level rise and climate.[25] While these examples do not dismiss the seductive allure of digitally designed data-visualizations, they credit the value of aesthetics in crafting and *elucidating* landscape conditions to have agency in the contemporary landscape.

Wind Screen
Cambridge, Massachusetts, 2011

Höweler + Yoon Architecture

Collaborators: Eric Höweler and Meejin Yoon, along with architecture students, visual arts students and mechanical engineering students

Scope: Temporary Installation on the MIT Campus

Technology: Miniature Wind Turbines self powering LED lights

Wind Screen exhibited among a series of temporary public installations for the Massachusetts Institute of Technology as a part of the *Fast Festival of Art, Science, and Technology* in celebration of MIT's 150th year. The research and architecture practice Höweler + Yoon Architecture was founded by partners Meejin Yoon and Eric Höweler who "believe in an embodied experience of architecture, seeing media as material and its effects as palpable elements of architectural speculation."[26] Their work has included a number of playful and imaginative responsive installations that both elucidate environmental phenomena and encourage interaction. As a faculty member at MIT, Meejin Yoon supported the installation of *Wind Screen*: an array of 500 wind-powered lanterns composed as an exterior screen on the facade of Building 54; as well as *Light Drift*: an interactive light installation of glowing sculptural seating elements and floating orbs along the Charles River, inspiring playful interaction between the work and the river's edge.

Wind Screen is a particularly interesting case study within their range of interactive and responsive experimentation for what Meejin describes as the project's "environmental-responsiveness."[27] Composed of 500 small-scale wind turbines or "wind lanterns" suspended by a steel cable armature, the lanterns self-power individual embedded LED lights to indicate wind energy in real-time. The collective performance of the miniature spinning wind turbines harnessing localized wind energy can be read as a diagram for the possible energy production of large-scale wind-energy infrastructures. Aside from highlighting the innovative possibilities of design and technology, the aim of the project is to change public opinion surrounding the resistance against the aesthetic impacts of wind power by coupling the function of a wind turbine with the aesthetic of an art object.

Many of Höweler + Yoon's responsive installations require human interaction to prompt and initiate responsivity, such as *Light Drift* in 2011 or *Swing Time* in 2014, a motion sensitive urban swing set of twenty illuminated ring-shaped swings that invite interaction and activate the space.[28] *Wind Screen* directly interfaces the phenomenon as it exists, whereas many other responsive installations require the

Figure 04.02 Wind Screen. Courtesy of Höweler + Yoon Architecture, Cambridge, Massachusetts, 2011

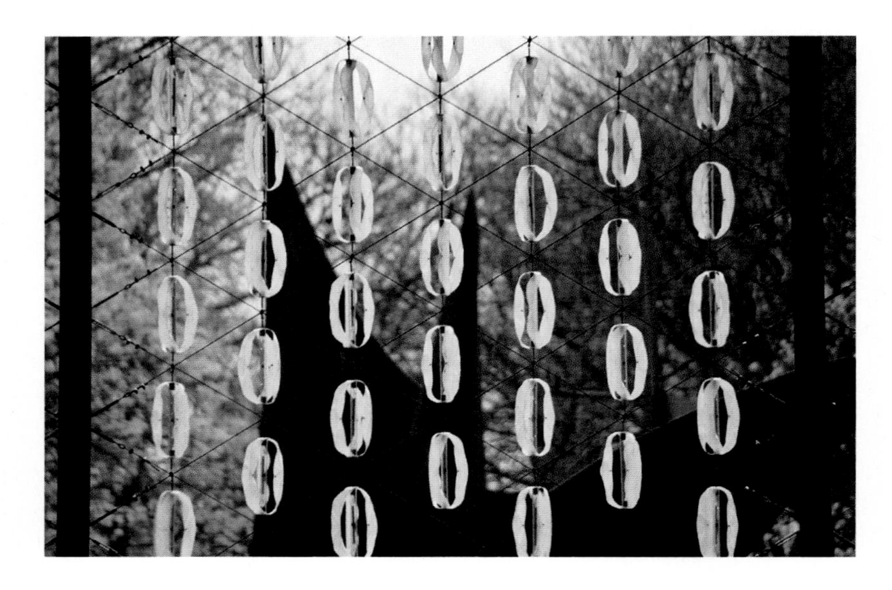

Figure 04.03 Wind Screen, *Courtesy of Höweler + Yoon Architecture, Cambridge, Massachusetts, 2011*

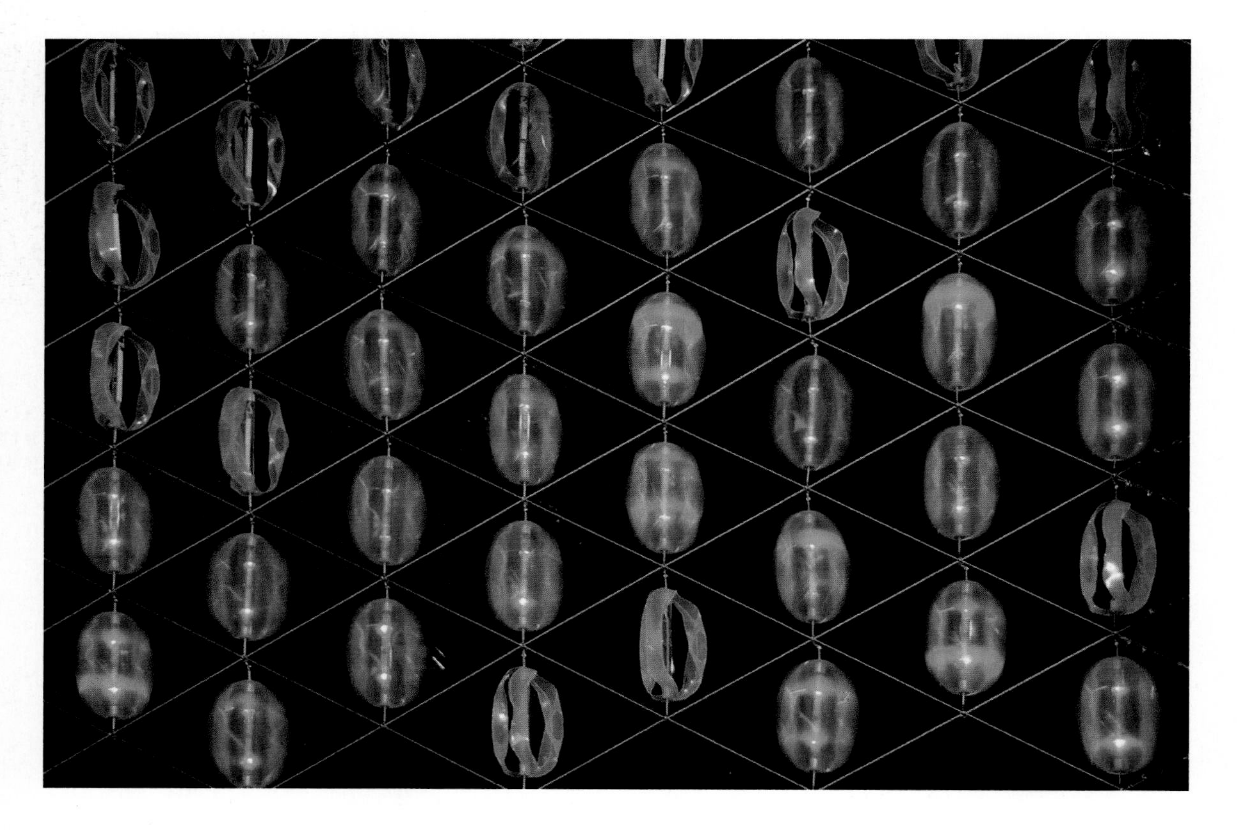

Figure 04.04 Wind Screen, *Courtesy of Höweler + Yoon Architecture, Cambridge, Massachusetts, 2011*

sensed information to be translated into a visual or experiential medium most often by computational methods. The input and sensory information required for *Wind Screen* is independent from human initiated interaction and only requires an energy input, in this case wind. The project not only indicates the presence of dynamic environmental phenomena through responsive technologies, but also elucidates the performative qualities of materials within the landscape, suggesting new possibilities for directly engaging with dynamic systems. In this sense, the engagement of external phenomena is less about catering to modes of interaction between human and object, and more about placing the phenomena initiating the response (wind) under investigation. In this case, the elucidation of wind power is tied to both the performance and sensory visualization through a one-to-one relationship in situ.

Figure 04.05
Wind Screen, *Courtesy of Höweler + Yoon Architecture, Cambridge, Massachusetts, 2011*

Figure 04.06 Dune 4.2, Studio Roosegaarde, Esch, Rotterdam, Netherlands, 2010. Photograph by Daan Roosegaarde.

Dune 4.2

Esch, Rotterdam, Netherlands, 2010

Studio Roosegaarde

Collaborators: Daan Roosegaarde

Scope: Permanent interactive landscape corridor of 60 meters across the Maas River in neighborhood 'Esch' Rotterdam, NL. Open for public at night

Technology: Hundreds of fibers, steel, sensors, speakers, software, and other media

---

Dune 4.2 is one installation in a series of iterative interactive landscapes designed by Studio Roosegaarde to produce a "new nature" evolving through technological innovation and spatially configured interfaces that interrupt and mediate architectural space. The installations in the *Dune* series are composed of the same "intelligent" prototype distributed within different spatial conditions: a simple and elegant branch made up of a thin and flexible straight black tube scaled to a tall grass or reed culminating in an encased LED light.

Figure 04.07 Dune 4.2*, Studio Roosegaarde, NIMK, Stedelijk, Amsterdam NL, 2006–2011. Photograph by Daan Roosegaarde*

Equipped with microphones and presence sensors, the fields of branches respond to disturbance in the form of noise, agitation, or movement by illuminating the LEDs. In each version of *Dune*, the landscape of fluctuating movement and light elucidates phenomena related to place as a combination of the viewer's energy within the space and the interaction between the branches, architecture, and surrounding systems. In Lucy Bullivant's essay "Alice in Technoland"—titled after Roosegaarde's phrase relating to the in-the-moment and

Figure 04.08 Dune 4.2, *Studio Roosegaarde, NIMK, Stedelijk, Amsterdam NL, 2006–2011. Photograph by Daan Roosegaarde*

supernatural experience after Lewis Carol's *Alice in Wonderland*—she describes Dune 4.0 as a "prosthetic" device employed as both an extension of the architecture and an extension of the participants' activities.[29]

The intelligent prosthetic acts as a mediator between public and private spaces, encouraging interesting social behaviors as the viewers learn and engage with the mode of response, testing the limitations and responsive effects. Daan Roosegaarde describes the iterative and prototyping process of the *Dune* series as a way of evolving the social and interactive environment: "We want it to learn how to behave and to become more sensitive towards the visitor."[30]

Dune has been reformatted to transform multiple spatial conditions—embedding and amassing the branches within the existing architecture creating unique landscapes. The massing, proximity, and sheer quantity of the individual branches has a distinct landscape feel—playing off the ephemeral qualities of vegetative fields to manifest sweeping winds and slight disturbances where small movements are amplified through the surrounding vegetation.

Dune 4.2 is a particularly interesting case study because of its context. The branches permanently span a sixty-meter bridge crossing the Mass River in Rotterdam. The immediate sensory actuation of the technology utilized renders the same method of response from the movement and noise of people to wind traveling across the water—horizontalizing the response to sounds and movement of both people

and environmental phenomena. Currents of light drift across the installation at night providing multiple perspectives of the installation to both active participants crossing the bridge and from afar, elucidating typically unseen phenomena. Depending on the intensity of the movement or sound, the response is scaled, resulting in surprising flashes and crashes or subtle glows and murmurs.

Figure 04.09 (facing page, top)
Dune 4.2, *Studio Roosegaarde, Esch, Rotterdam, Netherlands, 2010. Photograph by Daan Roosegaarde*

The sensitivity of the individual branches in such quantity and distribution emits the response as a field condition—elucidating dynamic and spatial gestures of sound and movement through a cumulative reading. In this sense, the method of elucidation supports current methodologies for interpreting and diagramming landscape scale readings of environmental phenomena. There are numerous ways to perceive or discern landscape systems through physical manifestations within the landscape. *Dune* speaks to the potential for using inherent landscape logics—in this case a field of tall grasses to render gusts of wind like a canvas—to extend our ability to translate and read invisible phenomena.

Figure 04.10 (facing page, foot)
Dune X, *Studio Roosegaarde, 18th Sydney Art Biennale, in Dogleg Tunnel at Cockatoo Island, Sydney, Australia, 2012. Photograph by Daan Roosegaarde*

Interview with Daan Roosegaarde

Dune and *Smart Highway*

Artist and innovator Daan Roosegaarde (1979) is internationally known for creating social designs that explore the relation between people, technology, and space. His Studio Roosegaarde is the social design lab with his team of designers and engineers based in the Netherlands and Shanghai. Roosegaarde is in the top five of Sustainable Trouw 100 as most Dutch influential green leader and selected as Talent of the Year 2015 by Kunstweek.

With projects ranging from fashion to architecture, his interactive designs such as *Dune*, *Intimacy*, and *Smart Highway* are tactile high-tech environments in which viewer and space become one. This connection, established between ideology and technology, results in what Roosegaarde calls 'techno-poetry'.

Roosegaarde has won the INDEX Design Award, World Technology Award, two Dutch Design Awards, Charlotte Köhler Award, Accenture Innovation Award, and China's Most Successful Design Award. He has been the focus of exhibitions at Rijksmuseum Amsterdam, Tate Modern, National Museum in Tokyo, Victoria and Albert Museum, and various public spaces in Rotterdam and Hong Kong.

Selected by Forbes and Good 100 as a creative changemaker, Daan Roosegaarde is a frequent invited lecturer at international conferences such as TED and Design Indaba, and TV media guest at De Wereld Draait Door and CNN.

01

Your work embraces the technological constraints of the moment, from the versioning of *Dune* to the speculation of Smart Highways, the projects are technologically advanced while remaining within the realm of constructability. What are some of the outcomes this grounded approach provides for Studio Roosegaarde beyond constructability?

We are in the mud, literally. This is a time to dream and to have radical ideas that are realized by a desire to build and to make it happen. The fun part about working within this current moment is that not only government, but industry and entrepreneurial companies have an ambition to be a part of this field. Since the old system is crashing, the new one is still unknown. The challenges of the future are providing space for collaboration and innovation. In this moment, I believe the role of the architect, the designer, and the artist is to claim that space.

02

Similar to product design and software development, Studio Roosegaarde uses versioning as a way to evolve projects, for example the multiple iterations of the *Dune* project. How does this process reflect software development and do you see it as a way to construct larger landscape and architectural projects?

The important aspect of this working methodology is not to copy through versioning, but to learn, fail, and evolve. Not only in terms of developing technologies, but expanding this method to developing modes of interaction that overlap with what I have in mind as the creator. Through my own observations, architecture in the physical world is seemingly becoming more and more generic over the last few years, cities are beginning to look the same. The real creative building is happening in the virtual world. As a collective society, we would rather share our emotions with Facebook or Twitter than express them in public spaces. The architect is currently meaningless and absent in virtual space. The question for the designer becomes: how can we merge the material world and the virtual world? *Dune*, in all of its versions, is about the merging of both hard and soft structures. The placement of the installation in pedestrian tunnels adds a poetic layer of technology. The addition of poetic layers of technology into raw urban spaces is something I am really excited about.

03

Your description of the poetic nature of *Dune* leads into a current conversation surrounding the integration of technology within urban environments. The smart cities movement claims that technology leads to advanced efficiency and intelligence of the city. The opposing view of this movement sees the efficiency and hyper control of technology as one-dimensional and reductionary. Your project *Smart Highway*, although branded as "smart" has a poetic or ambient approach to interfacing with infrastructure that bridges this discourse. Where do you see your work fitting into this discussion?

That depends on your definition of smart and whether it means becoming more machine or more human? Is smart city ruled by the IBMs of this world to reduce for human activity? Or it is it an enabler to explore and discover? Are we viewing this through the lens of George Orwell or Leonardo Da Vinci?

For me, "smart" is tied to creativity, smart is realizing what Marshall McLuhan once said, "There are no passengers on spaceship earth. We are all crew."[31] Making places that enable embodies my interpretation of smart. I would rather relate to creative cities instead of smart cities, where technology is the facilitator or enabler and not the leader.

04

What are the specific challenges of prototyping responsive objects? How does the prototyping process feed into the construction of the deliverable product?

Within our studio, there are separate initiatives composed of designers, architects, and researchers working on multiple projects at different stages. Our projects always stem from examining a topic that fascinates me. To realize a project, we respond to requests from clients ranging between a private collector to the Mayor of Beijing, to a city. We then find the overlap to meet the needs of the client and our current research. There are hundreds of prototypes we have developed over time without a context yet, waiting to find their habitats. Our process begins by collaborating with architects to make sketch proposals, leading into a pilot project, a mock-up, and then onto a specific design proposal. After we have a proposal, we find a partner to realize it. To make this process possible, we need designers who are capable of collaborating with different people from different fields; architects are quite good at this.

We work collaboratively with multiple design disciplines and a range of people in engineering, electrical engineering, and computer programming. The fun part about this collaboration is that the new generation of designers grew up with a Facebook account, they know how to program and work with technology. The gap between disciplines is becoming smaller and smaller. Everyone has been experiencing the integration of technology into our lives and observed the failures and successes.

In our current development, we are moving away from microchips and working more with biological systems. We are working on a large light installation for Central Station and collaborating with astronomers to understand the differentiation and breaking of light. We are in a shift right now. We are moving towards solutions with biotechnology rather than the ubiquitous application of sensors and lights that respond when you walk by.

05

What are some of the current technological barriers to realizing projects that are environmentally responsive and interactive?

The desire and the challenge is to make landscapes that are on one hand energy neutral and at the same time incredibly poetic, engaging people in an open dialogue. We are crushing what we know as the art world and prototyping the new one. We are scared and intrigued by this transition. Strangely, the current environmental crisis has forced us to be creative. There has never been a time in my life, not since I was five years old, where there has been a stronger desire for innovation. Five years ago, a project like *Smart Highway* would have never materialized. A project like this would have placed industry completely outside their comfort zone. Now, because of the cost of energy there is a desire to engage radical ideas.

In order to realize environmentally responsive and interactive projects, architects must move out of the advisory role we've been pushed into and become curious—the kind of curiosity seen in the work of Toyo Ito and The Metabolists. My goal, personally and for the profession, is to make space for curiosity in architecture and creativity in the new generation.

06

What are you most excited about in regards to the upcoming technologies and how they might impact your firm's work?

We are looking towards synthetic biologies to transform our work into the future. We don't want to just build stuff, we want to grow stuff. The project we are prototyping for Central Station is composed of light emitting trees that will replace the maintenance, wiring, and electricity bills of conventional street lighting. Instead, we are looking at how a firefly generates light to develop a synthetic system based off of biological capabilities. Within the explorations of these technologies, there is a lot of room to find personal connections to spaces. We should be more curious about these new spaces then we should actually build them. Right now we are only a 15-person studio so much more can be done.

I am also excited about a current project we are working on titled *Smog Free*, in which we are essentially building the largest electronic vacuum in the world right in the middle of Beijing City to create a smog free park, literally sucking up all the smog to create the cleanest air in the city. Instead of disposing of the collected smog, we are collecting it and fabricating rings where the smog takes the place of the diamond—representative of 1,000 cubic meters of clean air. Because smog is made of carbon and carbon under a lot of pressure is a diamond, it becomes an interesting play on materials as well as the value of clean air. We compress the smog just enough to crystallize and house it in a clear cube, designed to look like an architectural building block. This project came about at an interesting time because the mayor just launched a US$160 million campaign to make the city smog free, so suddenly we are in the smog business. It is radical architecture—an architecture of action—by removing the smog you create a bubble of clean air. In a way this idea is inspired by Buckminster Fuller: we are creating a place where people can literally *smell* the future. The project attaches itself to the urban environment in a much larger way. *Smog Free* is an important project for the future direction of my work and representative of the future of responsive landscapes.

Nuage Vert (Green Cloud)

Helsinki, Finland, 2008

Helen Evans and Heiko Hansen, HeHe

Collaborators: Martti Hyvönen, the Environmental Director of Helsinki Energy [Helsingin Energia]; Jussi Palola, researcher at Helsinki Energy's Electricity Network Company; Esa Räikkönen Laser Physics Department at the Helsinki University of Technology; Dodo, Finnish environmental activist group; Devalence, a Parisian design group

Scope: Temporary performance installation, to elucidate emissions from the Salmisaari power plant, Helsinki, Finland from 22–29 February, 2008 as part of *Pixelache Helsinki 08: Pixelache University*

Technology: Thermal Imaging Camera, Medical grade laser beam (up to 8.6 watts), unique Processing script

---

Nuage Vert, translated as "Green Cloud" is a temporary installation elucidating the relationship between energy consumption in Ruoholahti, the West Harbour area of Helsinki and the outputs of the local Salmisaari coal-fired power plant. Conceived by Helen Evans and Heiko Hansen, collectively known as HeHe, *Nuage Vert* is one concept in a series of environmental art installations and performances titled *Pollstream* centered around air pollution and man-made clouds to "aestheticize emissions and chemical toxins."[32] Evans and Hansen use their artworks to elucidate that "today's clouds are man-made" and the Earth's atmosphere is now a man-made "anthroposphere" concealing hazardous materials, products of chemical warfare, toxins, pollutants, and emissions from manufacturing and transportation.[33] The increasing invisibility and ephemerality of hazardous airborne material phenomena is compounded by "the complex filtering technologies that are added to thermodynamic devices" producing fine cancerous particulate dust, "perceived [only] through scientific measurement."[34]

With *Nuage Vert*, Hansen and Evans draw attention to the invisibility of hazardous pollutants and the ephemerality of man-made clouds by "giv[ing] form to the industrial cloud."[35] By projecting a laser emitted outline of neon green light in real-time onto the industrial cloud, the "invisible digital infrastructure"[36] performs as a "massive power meter"[37] illuminating the night sky to encourage citizen awareness and participation in reducing energy consumption. Sited at 155 meters in the air, at the very top of the Salmisaari coal-fired plant smoke stack, the green cloud is visible from 10 kilometers away. Helen Evans explains: "Turning a factory emission cloud green, inevitably, leads to questions being asked. It shifts the discourse about climate change and carbon emissions from abstract immaterial models based on the individual, to the tangible reality of urban life."[38] Calling attention to the physical and tangible outputs of uniquely material issues related to most landscape scale disturbances shows tremendous

Figure 04.11 Nuage Vert, *HeHe, Helsinki, 2008. Image by Mika Yrjölä*

Figure 04.12 Infographic, Nuage Vert, *HeHe, Helsinki, 2008*

agency for the potential of responsive landscapes to contextualize normally abstract figures of environmental impacts.

With *Nuage Vert*, Evans and Hansen comment on the ineffectiveness of conventional measures of electricity consumption by the single household marked every month by an electricity bill or by the concealed and forgotten electricity meter to adequately reflect and change consumption practices. By augmenting this reading to measure the cumulative consumption of electricity by the entire city in real-time, it becomes impossible for the city's inhabitants not to connect their individual and collective actions to the physical visualization of their outputs.

The project was initially sited in France on the northern edge of Paris to bring citizen attention to the relatively new incineration plant filling the skies of Saint-Ouen and Ivry-sur-Seine with conspicuous man-made clouds in close proximity and influence to HeHe. After multiple failed attempts to realize the project, Hansen and Evans approached the environmental director of Helsinki Energy and after three years of negotiations successfully acquired permissions from Helsinki Energy. The project hinged not only on the permission to project onto the plant's emissions, but on access to the real-time data stream of collective energy consumption.

Figure 04.13 Nuage Vert, *HeHe, Helsinki, 2008*

The realization of the project was a unique collaboration between fields within culture, science, industry, communication, and ecology.[39] In order to produce the visualization aspects of the project, HeHe collaborated with technological specialists in laser physics and computer science. The outline of the cloud was sourced by a thermal imaging infrared camera and input into real-time image processing software. Using a unique script in Processing, the thermographic image sensing was translated into a vector outline of the factory emission through blob detection.[40] The outline was re-projected back onto the emission using a medical grade laser beam coupled with a scanning system to effectively outline the vapor plume 155 meters up into the lower atmosphere.[41]

Reducing the visual impact of the man-made cloud into a defined outline deconstructs readable elements of the cloud's significance, applying measurable values to an otherwise ephemeral form. The installation efforts were coupled with a clever and effective social campaign led by Dodo, an environmental activist group, and Devalence, a Parisian design group, reaching the public through participatory and pop-culture influenced graphic forms of communication. This effort encouraged participation and literacy of the installation event, resulting in a reduction of 800 kVA to mark the finale of the project.

The visualization of the green cloud had a seemingly inverse relationship to the production of energy; as residents unplugged and scaled back on their electricity consumption, the vapor plume grew larger. This phenomenon is explained by the overall reduction in energy use and the subsequent increase in vapor outputs as the plant releases excess energy due to limited storage capacity. Within the processing of the thermographic imagery, the outline is scaled in relation to the real-time evaluation of energy consumption—bridging the difficulties of interpreting the visual significance of man-made clouds. The juxtaposition of the green outline and the vapor plume elucidates the complexity and invisible measures of energy consumption. The striking aesthetic of *Nuage Vert* engages participants to ask, "Could this green cloud be a toxic cloud or an emblem for the collective effort of the local community?"[42]

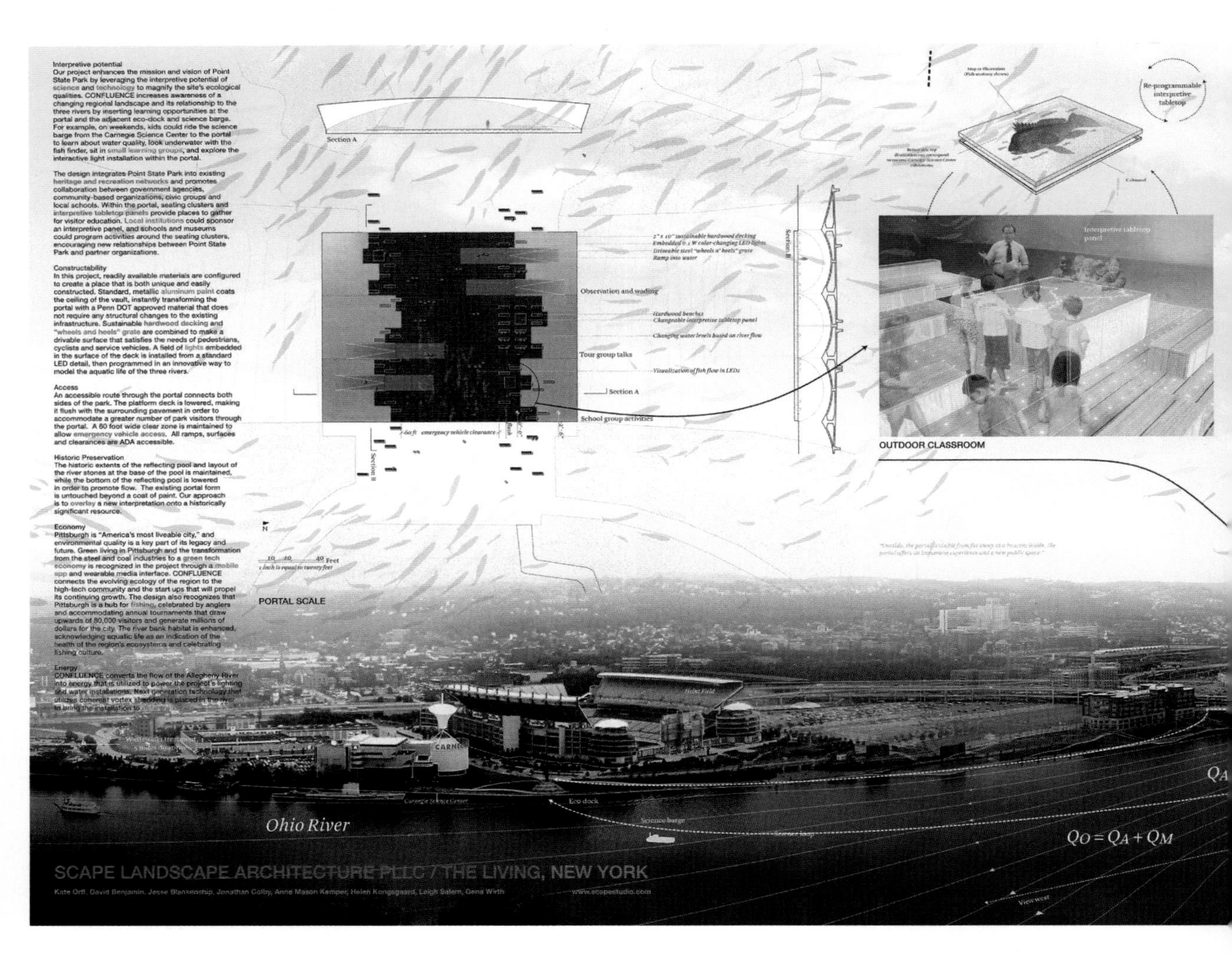

Figure 04.14 Competition board, SCAPE/LANDSCAPE ARCHITECTURE and The Living, New York, 2011

Science barge
Carnegie Science Center
Renewed fish habitat
Vortex-shedding technology
24 oscillating cylinders will produce an average of 1 kW, depending on river flow
Science loop
Science barge
Water quality monitoring
Water edge restoration
Fish monitor
Eco-dock
Water edge restoration
Fish monitor
Fish habitat restoration
Eco-dock above
Fish monitor
Vortex-shedding technology
Fish habitat
CE BARGE
ECO-DOCK
RIVER-BASED ENERGY GENERATION
Allegheny River
Monongahela River
Data visualization hub
River based energy generation
River flow monitoring
Phosphorescent paint
Eco-dock
Fish habitat restoration
Bass fishing hot spot
INPUT
River flow monitoring
INPUT
Water quality monitoring
Fort Pitt Bridge
Q_M

Confluence

New York City, 2011

SCAPE/LANDSCAPE ARCHITECTURE and The Living

Collaborators: SCAPE/LANDSCAPE ARCHITECTURE and The Living

Scope: Speculative Proposal for Pittsburgh's Point State Park

Technology: Water Flow Sensors (Flow) and LED Lighting Installation

---

Landscape phenomena is often obscured by the complexities of systems that are comprised of interconnected processes. The rising level of river water or a beautiful sunset can be seen as singular events and difficult to comprehend as temporal or atmospheric systems. The visualization of these ephemeral or invisible processes can help tie together how a site is situated within its context as a multitude of inputs and outputs the site engages. *Confluence*, a collaboration between SCAPE/LANDSCAPE ARCHITECTURE, a landscape architecture firm, and The Living, a design and research studio, illustrates these processes through both physical and virtual visualizations of environmental phenomena.

The project is located at Pittsburgh's Point State Park, at the confluence of the Allegheny and the Monongahela Rivers creating a portal by activating the underpass of the Duquesne Bridge. The project visualizes four major phenomena within the park; water quality, fish flow, people flow, and river flow. The separation and rendering

Figure 04.15
Learning landscape with indicators of river health, SCAPE/LANDSCAPE ARCHITECTURE and The Living, New York, 2011

Figure 04.16 Energy generation, SCAPE/LANDSCAPE ARCHITECTURE and The Living, New York, 2011

of each phenomenon creates discrete visualizations, bringing clarity to a complex series of overlapping events. The area under the Duquesne Bridge becomes the canvas for these visualizations through a light field at the ground level and the projection of light onto the base of the bridge structure. The portal from the city to the confluence animates a typically dark and possibly unsafe area of Point State Park through the visualization of fish within the nearby rivers.

The confluence of these two rivers is a known fish habitat for bass, and site for the annual Bassmasters Classic known as the "Super Bowl" of fishing tournaments. This fish habitat, while small for an industrial river system, is an important indicator of the river's health and a gauge for citizens to understand the river as living ecosystem. Fish are detected within the rivers through depth finders (sonar) and their presence is visualized within the plaza through sequenced LED lighting in the light field. This visualization iconifies the presence of fish in the river and renders them swimming within the plaza itself.

Water quality is rendered as a spectrum of color projected onto the underside of the bridge. This spectrum indicates the overall quality of the water, blurring the line between the occupation of the plaza and the river creating a visceral relationship between the two. As people move through the plaza their movements are mapped onto the projection of the water quality, blurring and altering the visualization into a composite visualization of human flows and water quality. This helps to heighten the relationship between residing in the plaza space and phenomena that occurs within the river.

Figure 04.17 Eco-barge and connection to Children's Science Museum, SCAPE/LANDSCAPE ARCHITECTURE and The Living, New York, 2011

Figure 04.18 Teaching cluster, SCAPE/LANDSCAPE ARCHITECTURE and The Living, New York, 2011

Physically the proposal also engages the river flow by rendering the water stages within a series of ramps that descend into the water. The slope of the ramps allow individuals to engage directly with the water and exaggerates the change in flood stages. Small changes in the river water levels are rendered on the gentle slopes of the ramps making it possible to observe large-scale river behaviors. Data collection on the site is concentrated in an eco-dock and science barge, which also function as outdoor classrooms. These spaces serve to consolidate the more ambient elements of the site into highly didactic experiences that are connected directly to quantitative studies of water quality and habitat. The designers also propose to produce electricity on-site through a series of oscillating cylinders that use the kinetic energy of the river for power generation.

It is through this combination of visualization techniques, didactic spaces, and on-site power generation that the project elucidates the complexity of the river system and the associated public space. The overlay of computational devices and physical visualizations blends multiple methods and proposes a highly diverse range of embedded technologies. In some sense *Confluence* renders a current or desired landscape condition, one that is not highly apparent, and then maintains a course of action. This creates a landscape that is highly stable and an ecology, habitat, and water quality that is tightly embedded to humans through methods of visualization. A heightened sense of connection between the vitality of the public realm and ecological fitness tends to herald a form of stewardship for these systems. *Confluence* attempts to stitch together this narrative and celebrate a form of interconnectedness that is readily apparent to casual occupants of the space.

NOTES

1 Malcolm McCullough, *Abstracting Craft: The Practiced Digital Hand* (Cambridge, MA: MIT Press, 1996), 43.

2 Ibid.

3 Interview with Hiroshi Ishii in Bill Moggridge, *Designing Interactions* (Cambridge, MA: The MIT Press, 2007), 527.

4 Pablo Valdivia y Alvarado, Vignesh Subramaniam, and Michael Triantafyllou, "Design of a Bio-Inspired Whisker Sensory for Underwater Applications," *Conference: Sensors*, 28–31 Oct 2012, doi:10.1109/ICSENS.2012.6411517, 28.

5 Ibid.

6 Dunne, *Hertzian Tales*, 109 (see chap. 1, n. 18).

7 June I. Medford and Ashok Prasad, "Plant Synthetic Biology Takes Root: Applying the basic principles of synthetic biology to plants shows progress," *Science* 346, no. 6206 (October 2014): 162.

8 Ibid.

9 See Chapter 7, "Connect."

10 Interview with Chris Woebken in Geoff Manaugh, *Landscape Futures: Instruments, Devices and Architectural Inventions* (Barcelona: Actar, 2013), 86.

11 Ibid.

12 Dunne, *Hertzian Tales,* 107.

13 Ralph Trancoso, Edson E. Sano, and Paulo R. Meneses, "The spectral changes of deforestation in the Brazilian tropical savanna," *Environmental Monitoring and Assessment: An International Journal Devoted to Progress in the Use of Monitoring Data in Assessing Environmental Risks to Man and the Environment* 187, no. 1 (2015): 1–15.

14 Suzanne Daley, "Greek wealth is found in many places, just not on tax forms" *New York Times*, May 1, 2010, New York Edition. http://nytimes.com/2010/05/02/world/europe/02evasion.html?_r=0.

15 DigitalGlobe is a private and leading global imagery company which provides high-resolution Earth imagery products and services by way of private satellite network. https://digitalglobe.com. Accessed April 10, 2015.

16 Satellite Sentinel Project, "Documenting the Crisis," http://satsentinel.org/documenting-the-crisis. Accessed April 10, 2015.

17 Linda Weintraub, *To Life!*, 2012 (see chap. 1, no. 21).

18 Scott, *Seeing Like a State* (see chap. 2, n. 12).

19 Manaugh, *Landscape Futures*, 25.

20 Ibid.

21 Dunne, *Hertzian Tales*, 101.

22 Ibid, 102.

23 Shannon Mattern, "Methodolatry and the Art of Measure," *Places Journal* (November 2013). Accessed 13 Feb 2015. <https://placesjournal.org/article/methodolatry-and-the-art-of-measure/>

24 See Ignasi de Solà-Morales, "Terrain Vague," in *Terrain Vague: Interstices at the Edge of the Pale* 24–30. (New York: Routledge, 2013), Eds. Manuela Mariana and Patrick Barron and David E. Nye, American Technological Sublime (Cambridge, MA: MIT Press, 1994).

25 Illustrated in Kristina Hill's exhibit *The Aesthetics of Necessity: Climate Change, Infrastructure and the Sublime*, The Wurster Gallery, College of Environmental Design, University of California Berkeley, February 7, 2013.

26 Website Firm Description: http://hyarchitecture.com/firm.

27 Eric Höweler and Meejin Yoon, "Wind Screen," Höweler + Yoon Architecture (blog), 2011. http://hyarchitecture.com/projects/48.

28 Eric Höweler and Meejin Yoon, "Swing Time," Höweler + Yoon Architecture (blog), 2014. http://hyarchitecture.com/projects/114.

29 Lucy Bullivant quotes Daan Roosegaarde in "Alice in Technoland," *Architectural Design* 77, no. 4 (2007): 7.

30 Ibid.

31 Marshall McLuhan, "The Relation of the Environment to the Anti-Environment," in *Marshall McLuhan—Unbound*, 4th ed. (Corte Madera, CA: Ginko Press, 2005), 17.

32 Helen Evans and Heiko Hansen, http://hehe.org.free.fr/hehe/pollstream/.

33 Helen Evans and Heiko Hansen, "On the Modification of Man-Made Clouds: The Factory Cloud," *Leonardo* 47, no. 1 (2014): 72–73.

34 Ibid., 72.

35 Ibid.

36 Helen Evans, "Nuage Vert," *Cluster Magazine* 07, (May 2008): 30–35.

37 Weintraub, *To Life!*, 206 (see chap. 1, n. 21).

38 Evans, "Nuage Vert," 30.

39 Ibid., 33.

40 Helen Evans, Heiko Hansen, and Joey Hagedorn, "Artful Media: Nuage Vert," *IEEE Multimedia* 16, no. 3 (2009): 13–15.

41 Ibid.

42 Helen Evans, quoted in Weintraub, *To Life!*.

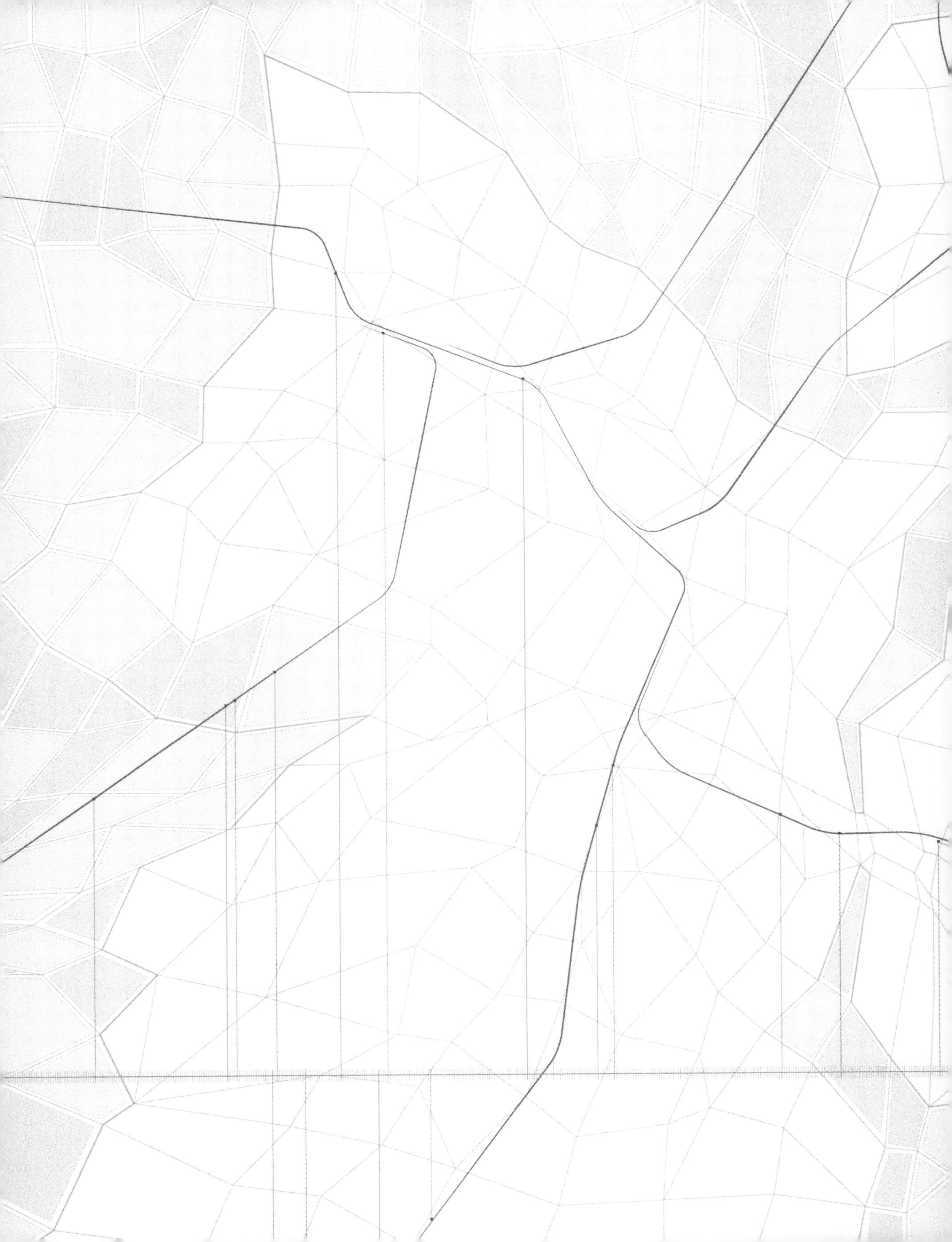

COMPRESS

05

The investigation and manipulation of temporal scales to interpret and decipher change over time, is the basis of *compress*. *Compression* augments phenomena and interaction by translating behaviors in time. This new arrangement alters known relationships and reinterprets them in a new temporal context. Long periods of time in which change occurs at an incredibly slow pace are particularly difficult to comprehend; therefore, *compression* re-makes and contracts these narratives into understandable scales of time. The *compressed* rendering of time becomes the lens through which we understand an evolved or dynamic landscape process. Interpretation typically occurs over timescales of human interaction by compressing time into digestible chunks. For example, a geologic timescale is made perceptible over a timescale of ten minutes. This interaction timescale is designed within a space, device, or interface to reconfigure our perception of events beyond our sensory scope.

The landscape illustrates the passing of time through daily solar and lunar cycles, seasonal change, growth, decay, and geomorphology. The metric of time, as a human construct, is utilized to make basic and fundamental decisions surrounding the organization and structure of behaviors. Once the United States national time standard was adopted in 1883 to co-ordinate railroads with common timetables, time was separated from local municipalities and local landscape conditions. Time remains the most fundamental metric utilized in understanding landscape processes, in the same way that the landscape itself records time. In his chapter, "Reclaiming Space," Alan Berger deconstructs the word "landscape" in which the combination of "land" and "scape" together are "defined as a temporal composition of man-made spaces on the land, not static natural scenery."[1] He further illustrates the connection between conceptions of landscape to manipulated temporal processes by quoting writer John Brinckerhoff Jackson:

> A landscape is . . . a man-made system functioning and evolving not according to natural laws but to serve a community . . . A landscape is thus a space deliberately created to speed up or slow down the process of nature.[2]

While the concept of time may have emerged from cyclic landscape processes, the co-ordinated, highly mechanized, and controlled global time is disconnected from localized landscape processes. James Carey describes this cultural shift in his text *Communication as Culture*:

> Today, computer time, computer space, and computer memory, notions we dimly understand, are reworking practical consciousness coordinating and controlling life in what we glibly call the postindustrial society.[3]

Time as a standardized metric is most valuable for connecting and co-ordinating information and actions. Time as a "mechanism of control," particularly within commerce and politics, has dramatically reshaped operations, extending control and co-ordination across global scales. In this sense the accuracy of time has become incredibly important and influential, to the point in which time cannot even be experienced, only envisioned in the abstract: "Time has been redefined as an ecological niche to be filled down to the microsecond, nanosecond, and picosecond."[4] The operation of these measures are tied to a complex and physical "network of networks," constituting the physical infrastructure of the internet. Composed of tubes holding fiber optics that span continents, this vast infrastructure determines where and with what speed information travels.[5] In his investigations of Los Angeles or *The Infrastructural City*, Kazys Varnelis researches the architectural artifacts of these conditions. The most valuable real estate in Los Angeles happens to be One Wilshire: "the prime communication hub between Asia and the Western world"—visually marked by "the world's most spray-painted asphalt" and covering the tremendous amount of fiber optic cords below that offers the fastest methods of communication across oceans.[6] This level of abstraction through mechanized time further separates qualities of landscape change in the context of human experience, in which "[p]ublic life is now scheduled and allocated more by time than centered according to place."[7]

Despite the global co-ordination of time, there are still varying cultural conceptions that insert slight differences into perceptions and control. "Mono-chronic" societies perceive time as stable and fixed with actions that occur sequentially, while "polychronic" societies consider time to be more malleable and open to dynamic occurrences. Another cultural distinction exists between the West and the East. Broadly,

Western cultures consider time to be linear (moving in a straight line), whereas Eastern cultures envision themselves as simultaneously existing in the past, present, and future because there is no distinction between past and future tense. In this sense, there is still a conception of time that exists beyond our metrics to link past, present, and future states. Although we may not be using traditional measures of landscape change to "tell time," the passing of time emerges across the landscape as it intersects with human timescales. For instance, daily, seasonal, yearly, geological, and ecological timescales remain distinct from urban timescales. The intersection of the two creates dependencies, disruptions, and augmentations in which alterations to the material qualities of the landscape continue to engage with dynamic landscape processes—whether accelerated, interrupted, or disrupted, matter remains continuously acting and "vital."[8]

With the measure of time as a constant factor, time becomes the metric in which other relationships are drawn out. Time is perhaps the most valuable metric for understanding dynamic landscape conditions. Linda Weintraub asserts that "anything occupying space also transforms through time." Her research into the genre of contemporary eco-art notes the relevance of relating artistic conceptions of dynamic relationships to a particular variable of time: "While flux is inevitable, its tempo is variable, determined by the inherent responsiveness of a medium and the intensity of surrounding influences."[9] Specific dynamic processes may tell completely different narratives between micro- and macro-timescales. Micro-timescales may be concerned with seconds and fractions of seconds; conversely, macro timescales may be concerned with geologic or deep time. Between these extremes there is an infinite range of tempos or velocities of time scaled in relation to human and contextual timescales. This variable can be determined by understanding change—both chemical and physical—over time through cycles of growth, decay, or evolution, in which the subject of compression plays a role. From the molecular to the geologic, the captured processes will produce distinct narratives through varying rates of compression.

Change occurring across human timescales is rapidly advancing. Current landscape dynamics can no longer be defined by a rate, but rather a velocity or speed,[10] in which movement of material, energy consumption, carbon emissions, and population growth all take the form of asymptotic curves. In general, *movement* is rapidly changing perceptions of the landscape in relation to time. There is the movement of people in the form of travel, making it less likely for a person to stay in one place and observe changes throughout a lifespan. Then there is the movement of material, drastically changing the way processes occur. Movement and transportation remove the limitations of time, drastically altering the timing or velocity of landscape related processes. For example, most fruits and vegetables are available at any time of the year.

Revealing change over time is achieved by elucidating previous conditions, revealing geologic processes, or highlighting material manifestations. The velocity or speed of time, at the scale of landscape change, is mostly imperceptible and requires intimate knowledge of a place, so much so that the notion of stability has been assigned to inherently unstable ecosystem conditions. In *Projective Ecologies*, Chris Reed and Nina-Marie Lister describe the "profound implications from the humanities to management applications to design" as these fields work to undo the misconception of ecosystem states as stable, shifting perceptions that rely on "the recognition that humans are not outsiders to the ecosystem—rather, we are participants in its unfolding."[11] At a human scale, there are few times when landscape change is obvious due to the "human, time-limited perception of stasis."[12] Change is most noticeable when there is a known datum. Rising water levels in coastal conditions are revealed by static (built) structures as the landscape changes around them. Natural disasters, particularly those heightened by failures of infrastructure, are moments when dramatic landscape change occurs within human timescales.

Understanding the rate of ecological change is critical for the implementation of large-scale ecological management strategies and plans. There is now a substantial investment in sensor networks and open access to data. Mandated by the US Constitution to proportionately assign seats in the House of Representatives, the United States Census Bureau has been conducting the census, an official count of the population, every ten years since 1790.[13] The basic questions in the census when related to geographical location provide an incredible resource for understanding urban and landscape conditions over time. The time-scale slider in Google Earth is an incredibly accessible tool for easily and quickly observing landscape scale change over time. However, the scope is limited to the collection of aerial imagery. There are a number of larger governmental institutions of landscape monitoring that relate acts of measurement to intervals of time. Satellite functions in co-ordination with land-based static monitoring stations become markers throughout the landscape, consistently measuring the landscape at specified intervals of time. For surveying at the planetary scale, there is the Landsat network. The National Agricultural Imagery Program is administered by the US Department of Agriculture's Farm Service Agency (FSA) to acquire high-resolution orthoimagery during grow seasons, initially conducted every five years, and now every three.[14] When coupled with the agricultural census taken every four years, this data presents interesting studies over time revealing agricultural shifts across the United States. These examples of compression can only extend so far back in time and rely on the institution of the sensing devices as well as the interval of collection to illustrate change.

At the human scale, current technologies such as smart phones continuously collect data through embedded sensors furthered by trending personal monitoring applications. These applications record habits of sleep, exercise, and movement patterns—rendering them, through infographics, to reveal health data over time. The vast quantity of data collected through different monitoring technologies only becomes important when it is distilled. A method of recording and playback, in which information or landscape processes are connected to time, provide the components to establish and shift velocity.

Compression operates through the acceleration and deceleration of time-based velocities using familiar techniques such as time lapse. This form of compression is apparent in photographic techniques such as Harold Edergton's experiments in stroboscopic and rapatronic photography.[15] Edgerton's technique of flashing a strobe to capture aggregate movements in a single negative or to capture film at up to 18,000 frames per second revealed a world that was outside of human senses. This form of exposure could catch the moment a bullet sheared through the edge of a card or the splatter of a drop of milk. These views into time—unpacking the millisecond so that it can be observed, studied, and analysed—extends human cognition and creates an entry point to develop devices that manipulate this temporal scale.

As a discipline, landscape architects utilize representations and visualizations that project into the future, typically around fifty years to exhibit mature and robust vegetation. This trend is problematic, particularly as the discipline shifts towards time-based ecological processes. Compression is a powerful tool for facilitating narrative and the sequencing of events, for obvious didactic qualities as well as a tool for altering perception. Narratives operating through compressed timescales speak to present, past, and future conditions, curating the time frame in which dynamic relationships are perceived, ultimately making a statement or privileging particular processes. Compression may also utilize time as a comparative tool, requiring something else to hold still, with use of datums. Similarly, the methods in which time is utilized to interpret change alters approaches to manipulations and operations within the landscape.

The coupling of geo-referencing and time-based landscape sensing, provide opportunities to understand the landscape through different timescales, recognize trends and make predictions. This has applications and implications for how we manipulate landscape processes, ultimately informing decision making. "Information technology causes profound changes in the time-frame patterns of the decision-making process."[16] Previously, the decision-making process was based on previous experience or recorded scenarios, with heightened technological tools for simulation and production of virtual realities. "We can gain experience of scenarios or events that have never been encountered in real life" in which the "simulation of the future . . . thereby modifies the time frames which are no longer relegated to repetitions of the past with little variation."[17]

The ability to compress (or expand) time conceptualizes a continuum, a navigable terrain that can be explored to understand a system's temporal relationship. This continuum may be virtual or physical, simulated or material, in which virtual models attempt to simulate processes across time. Physically, through generations of monitoring, this continuum exists as an observation or abstraction of ecological systems. Data that is gathered and reapplied to interpret histories is projected and manipulated to understand or find new futures. This form of compression is common in most simulations and forms of forecasting such as ensemble forecasting. Compression can be a valuable tool to address the "temporal shift" in ecological theory, which considers old conceptions of bounded sites as "part of a changing context in which trends cannot be exactly predicted, and surprises should be expected." These changes cannot be determined solely by a velocity, but rather, recomposed as a narrative to extract meaning across multiple scales of time.

Figure 05.02 Prototype of The Theater of Lost Species, *Future Cities Lab, San Francisco, California, 2014*

Nataly Gattegno and Jason Kelly Johnson, Future Cities Lab

Collaborators: Ripon DeLeon (Senior Associate), Shawn Komlos (Intern)

Production Team: Ji Ahn, Fernando Amenedo

Fabrication: MACHINIC Digital Fabrication & Consulting, San Francisco

Scientific Collaborators: Matthew Clapham (UCSC), Dr Jonathan Payne (Stanford University)

Scope: Interactive installation for the exhibit at the YBCA "Dissident Futures" show Oct 18-Feb 2, 2014

Technology: Integrated IR Sensors, Arduino microcontrollers, Python, Ruby, Processing, Grasshopper, Firefly, LEDs; Digital Fabrication: CNC Milling, 3D Printing, and Laser Cutting

---

As we augment and anti-terraform[18] the Earth to become both uninhabitable and unrecognizable between generations, how do we continue to know and to mourn the species and landscapes no longer present, and possibly never known? *The Theater of Lost Species* designed by Future Cities Lab constructs a narrative around species loss and global mass extinctions in response to a timely Op-Ed piece in the *New York Times* by Lydia Millet, asking the question of future generations who will undoubtedly reside in an "alien" landscape: "can you feel the loss of something you never knew in the first place?"[19] Exhibited in San Francisco at the Yerba Buena Contemporary Art Center as part of the *Dissident Futures* exhibit, *The Theater of Lost Species* performs as a prototype for virtualizing extinct species and landscapes (in this case extinct sea creatures) through an interactive digital interface.

Standing 16 feet tall and 16 feet in diameter on robotic legs, the *Theater* as a collector and repository of landscape information looks somewhat like a Mars Rover equipped to autonomously navigate and explore unknown and uninhabitable territories—predictive of a new future cyborgian species, this is a creature to record and remember former creatures. The current prototype is designed more as a "theater" than as a "collector"—meant to responsively engage viewers to explore the virtual lost species housed within. The *Theater* is encased in digitally fabricated Fiber Reinforced Panels and resin, forming viewing cones into the interior and aligning views to "digital display screens that are portals to a seamless virtual aquarium."[20] Each viewing cone is equipped with three integrated IR sensors to sense proximity. As the viewer peers into the *Theater*, it actuates a custom physical–digital interface to display species. At the same time, fiber optic plastic "hairs" illuminate on the exterior, calling attention to the

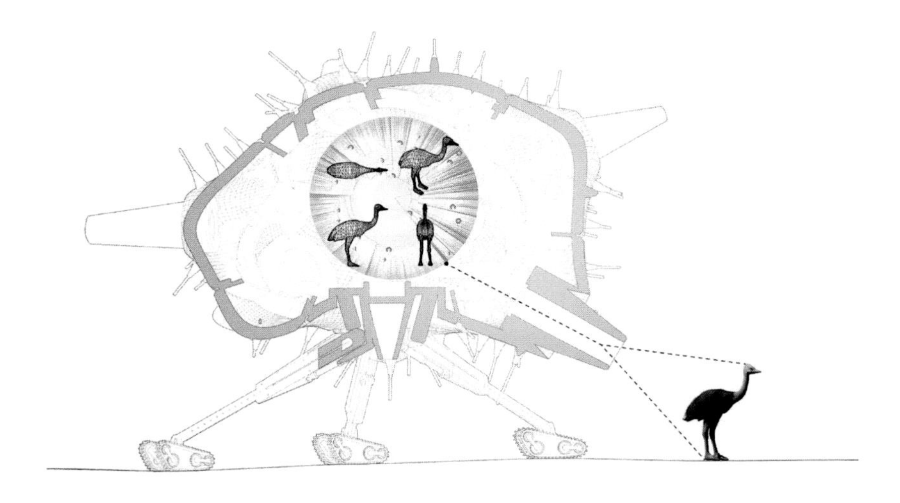

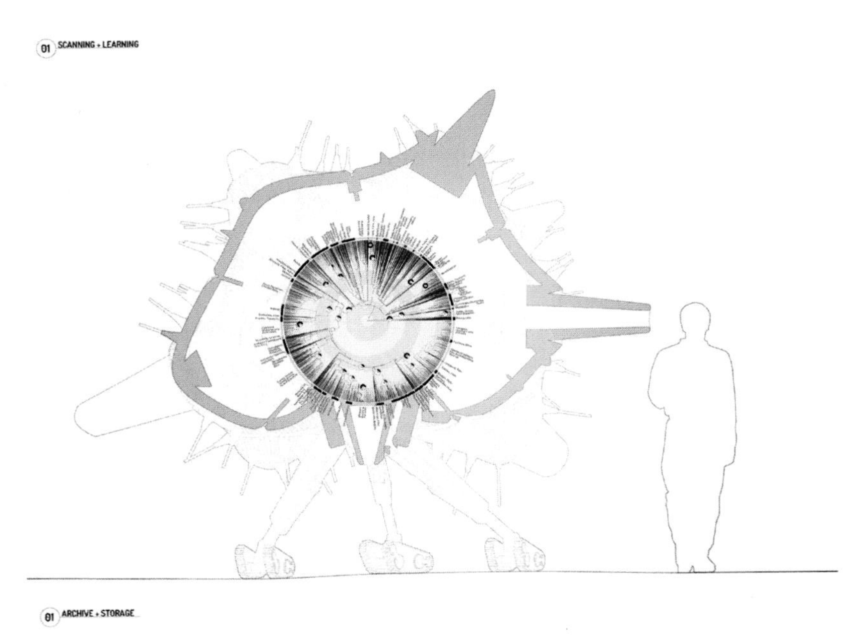

Figure 05.03 *Scanning, learning and archive, and storage design diagrams*, The Theater of Lost Species, *Future Cities Lab, San Francisco, California, 2014*

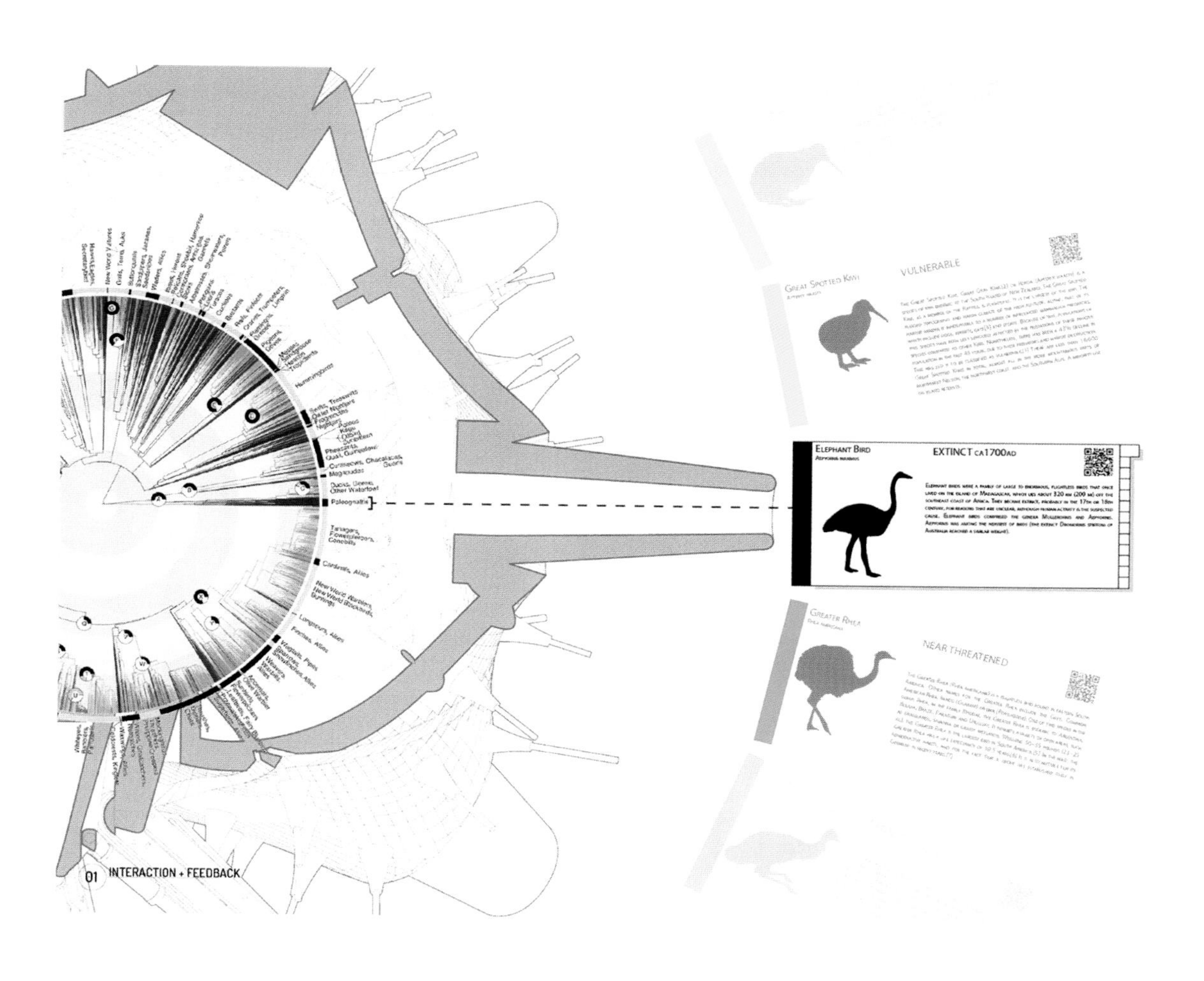

Figure 05.04 Interaction and feedback design diagram, The Theater of Lost Species, *Future Cities Lab, San Francisco, California, 2014*

previous life forms housed within. The interactive components were programmed through the combination of Processing, Arduino, Python, and Ruby.

As we collectively speed up geologic change, our current methods for recording, visualizing, or memorializing past landscapes will become increasingly valued as both historical record and as tools to inform landscape management practices within new ecological conditions. These records and representations will perform as both informative data and a cautionary tale as we continue to augment the landscape. A significant feature to the compression or manipulation of timescales is the ability to see large shifts that may span long periods. This is particularly true for timescales incorporating geologic or evolutionary change in which it would be otherwise difficult to connect instrumentality with outcome. Concepts explored within *The Theater of Lost Species* offer insight into the re-telling and re-formation of environmental augmentation over time.

Figure 05.05
Conceptual rendering,
The Theater of Lost Species,
Future Cities Lab, San Francisco, California, 2014

Although *The Theater of Lost Species* is designed to convey a narrative about extinct species, this idea of using a device to play-back or preserve past conditions of the landscape is quite relevant to the concept of "compress," which requires rendering of often imperceptible timescales. Establishing sensors and more devices to capture or collect data about the landscape will build a substantial catalog of previous conditions. The question then becomes: what do we do with all this data and how do we curate, organize, and access it? Inspired by similar forms of predictive collection and memorial-ization in anticipation of future change and loss, Future Cities Lab was inspired by Traveling Menageries, Chinese Lanterns and portable Camera Obscura devices developed in the 1800s, time capsules in the 1950s and 1960s, and the recent seedbank in Svalbard, Norway.[21] Though *The Theater of Lost Species* exhibits the extreme condition of extinction, it is predictive of the kinds of tools and devices that will help to memorialize past landscapes in order to navigate future landscapes.

Figure 05.06 Real-time thermal imaging projection, Emergence, *Obscura Digital*, San Francisco Exploratorium, California, 2013

STORE
AT PIERS 15/17

Emergence

San Francisco, California, 2013

Obscura Digital

Collaborators: Garth Williams (Creative Director)

Scope: Temporary live projection mapping composed of real-time and pre-recorded staged film pieces

Technology: Architectural Mapping, Augmented Reality, Motion Graphics, Projection

---

Emergence is a temporary projection installation on the exterior of the San Francisco Exploratorium designed for opening day by Obscura Digital to highlight the dynamic biological systems contained within the Exploratorium. The team produced all the imagery from physical analog phenomena both performed and live then manipulated the scale and time interval of multiple systems to reveal the interconnectedness of biological systems. Contrary to many live map projections that use digitally rendered, simulated, or enhanced imagery, the complexity embedded in the authenticity of physical materials in this piece inspires wonder and curiosity.

Figure 05.07
Cloud tank projection,
Emergence, *Obscura Digital,*
San Francisco Exploratorium,
California, 2013

The footage for the projections was developed in a studio using multiple-scaled analog models of the building façade to capture physical and biological dynamics at multiple scales over varying periods of time. The footage was then edited and projected onto the actual façade of the building during opening night, shifting and oscillating the velocity of the film to highlight time as a defining factor in the performance and perception of dynamic systems. Concurrently, thermal cameras captured real-time imagery of visitors to project alongside the composite of dynamic systems, placing the viewer within the compressed timeline.

Figure 05.08 The making of wax melt liquefaction, Emergence, *Obscura Digital, San Francisco Exploratorium, California, 2013*

The physical models were built to several scales, from microscopy to a four-foot water tank, designed to facilitate scalar material relationships and processes. Building scaled physical models and filming the real physical processes—instead of rendering a digitally modeled particle flow simulation—captures authentic reactions of the materials to the detailing and proportions of the building. The modeling efforts explored both biotic and abiotic processes. The miniature sets staged "fluid dynamics, wax liquefaction, particle interactions, microorganisms, living systems, crystallization, and growth in time lapse."[22]

Elucidating the physical abiotic processes required a range of materials. The four-foot water tank was designed with 36 individual valves for injecting multi-colored fluids to observe the entropy of dye moving and infiltrating. The model was filmed upside-down and reprojected onto the building making the fluids appear to defy gravity.

Figure 05.09 Protozoa microscopy projection, Emergence, *Obscura Digital, San Francisco Exploratorium, California, 2013*

Multiple colors of wax were melted and merged together across a model with a heat gun, exhibiting a slower and thicker mixing of materials. Another model was used to capture particle interactions materialized through sand directed by air flow manipulations. Fluid dynamics and material relationships were explored at a smaller scale through the mixture of pigments, water, and oil shifting in rejection and acceptance.

The colorful physical processes were coupled with biological systems at multiple scales. The largest scale recreated a terrestrial biopsy of a local Northern California forest floor composed of soils, vegetation, and living organisms at a tangible scale filmed over time lapse to capture growth and exaggerate material and energy flows. Explored at the microscopic scale, the designers etched the scaled facade into a microscope slide to film the Physarum life cycle—a slime mold found within decaying organic matter on forest floors in shade and moisture —filmed close enough to see the slime mold eating microbes such as fungal spores and bacteria. Projected onto the building facade, the microscopic scale showed the organisms jittering between the windows and moving across the building.

To situate the viewer as part of the environmental phenomena and further relate temporal scales within the compressed depiction of dynamic processes, a series of thermal cameras captured the visitors' motions in similarly bright fluctuating colors across the facade. The color palettes for the thermal imaging heat maps were sampled from

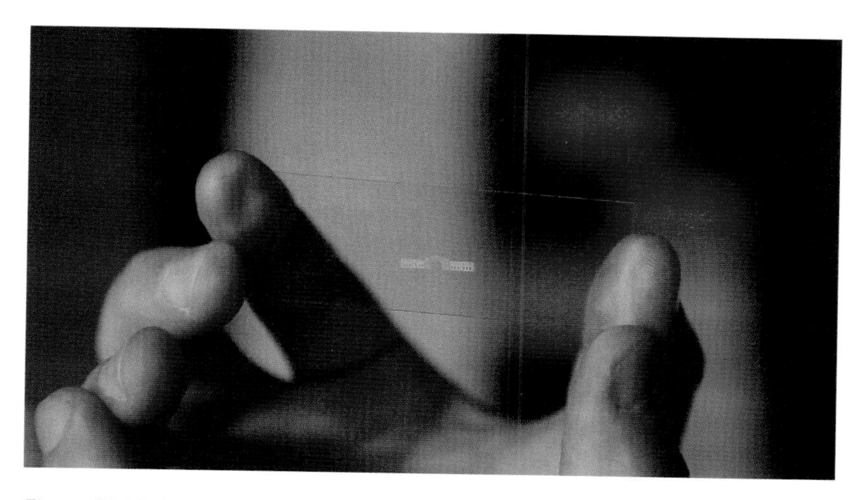

Figure 05.10 Protozoa microscopy microscope slide, Emergence, *Obscura Digital, San Francisco Exploratorium, California, 2013*

the footage of physical and biological systems—further correlating and placing human activity within the series of interconnected systems. *Emergence* prompts the viewer to question the broader context, positioning oneself within this incredibly large and dynamic system.[23] The goal of this project was not to elucidate temporal environmental phenomena but to display connections and relationships across physical and temporal scales, soliciting the viewer to contextualize "the nature of being" as they enter the Exploratorium.

Interview with Garth Williams

Emergence

Born in Pennsylvania and raised in Tennessee, Garth Williams received a degree in non-objective / conceptual Art from the Savannah College of Art and Design in 1992. Upon graduating, he worked odd jobs while pursuing animation and programming in service of exploring new frontiers in Art. By 2004, these pursuits lead him to AKQA San Francisco where he served as Senior Developer, Senior Creative and Innovation Director over the span of six years.

Shifting careers in 2010, Garth joined IDEO as a designer, specializing in concept and interaction where he collaborated with the US Intelligence Community, the Consumer Financial Protection Bureau and other entities. In 2012 Garth joined Obscura Digital as Creative Director where he conceived and designed numerous experiences for clients such as ATT/Dallas Cowboys, The San Francisco Exploratorium, the America's Cup and Genentech. After leaving Obscura in 2015, Garth became the Director of Innovation at The California Academy of Sciences.

01

Most live map projection is digitally rendered and simulated. For Emergence, the projections were composed entirely from live capture of physical and biological materials. What were the opportunities and limitations of this method? How did pre-production and production differ from other projects?

The intention of this work was to create a genuine connection to the primeval forces that define and shape the reality we share. To express this, computer graphics would never do. Living in an age of pseudo-experience, our senses are deceived so frequently that experiences created through simulations, facsimiles, and abstractions (digital) have superficial meaning and little emotional impact. Anything less than total authenticity would fall short of what the Exploratorium represents.

As for limitations, each of the different pieces in the show largely arose from constraints of time and space, context, and concept. Depending on one's perspective, limitations can be powerful allies, opportunities as opposed to obstructions. I believe the greatest work comes from a mindful collaboration with cosmic circumstance.

02

The scenes of physical and biological materials all have implications of scale. How did the role of scale facilitate the overall narrative of the performance? What decisions were made in the translation from physical process to projection?

Scale, both spatial and temporal, were very challenging. Each of the phenomena captured had its own specific requirements. The plant time-lapse piece required thirty days and specific physical dimensions to ensure that there were both enough time and space to fully capture the process of growth and interaction with architectural features. Most challenging was the microscopy piece: it required a whole new way of thinking (microns rather than feet or inches). The intention was to make it clear to the viewer that the protozoa were contained within the confines of the architecture, and ideally that they be able to traverse the spaces between the windows. To achieve this took a good deal of research about protozoa, ultimately leading to the magic number of approximately 300 microns, which is the median value of the protozoa we hoped to capture. Once this was established, over 20 different variety of protozoa were 'auditioned', for a variety of criteria (motility, opacity, color, and scale).

As for the translation of these experiences to the projected image, we did our best to contain, execute, light, and film for the most impactful and high-resolution image possible. We used a 'Red Epic' camera for many of the pieces, some of which were shot and assembled from multiple plates. The time-lapse pieces were done with multiple Canon 5D SLR cameras, once again shot and assembled from multiple plates.

03

In addition to the pre-recorded film, there was simultaneously a real-time live capture of the visitors, scaled and mapped to the building with a "thermal" filter. What technology was used? How did it facilitate interaction and change the viewer's perception?

To create the "Thermal Mirror," we arranged to have two thermal imaging cameras loaned to us from a company called FLIR (from a technical term: Forward, Looking, InfraRed). The interaction was simple: people stood in front of the camera and saw the heat signature of their bodies projected on the wall.

04

What amount of the performance was live or responsive to on-site phenomena versus predetermined? How were the methods and footage integrated?

The only part of the performance that was real-time was the thermal imaging wall. All the other pieces, though totally authentic, were carefully planned and executed studio shots. With more time and budget, many of these experiences would have been interactive and most likely real-time.

05

After the initial filming and live capture of the footage, there was a significant amount of post-processing to augment the velocities, compressing and expanding the timescales of the dynamic processes. Can you elaborate on the process of editing, combining footage, and coupling of processes?

Actually there wasn't too much post-production with regards to timescale. More dramatic effects, like the time lapse of the crystal and plant growth, were achieved in camera through timed exposures. However, there was one shot where we combined multiple cloud tank shots and compressed the timescale, which gave it a more dynamic, explosive look. As for the nitty-gritty editorial aspects of the media, it was extremely unwieldy and time consuming. The plates were huge, the microscopy being a whopping 18k. Even with a moderate rendering farm at our disposal, it put us in a tenuous place with deadlines and extreme sleep deprivation.

06

Live map projection has clear applications and overlaps with interactive design and augmented realities to shape the way we perceive and interact with the built environment. How do you see the application of live mapping and associated technologies evolving in the future?

Hopefully it will replace some of the morbid dependencies our culture has on mobile experiences. In some ways I'm optimistic about the utilitarian, experiential, and artistic possibilities of embedded and mapping technologies. On the other hand these technologies could usher in a dystopian nightmare of large-scale, immersive, and interactive advertising perversions. Presently, our minds are being polluted with useless and corrosive information at unprecedented levels. At the same time our most valuable cognitive resource—attention—is being looted by the latest machinations of 'Branded Experiences'. It will be interesting to see how these new methods impact our culture and the evolution of consciousness.

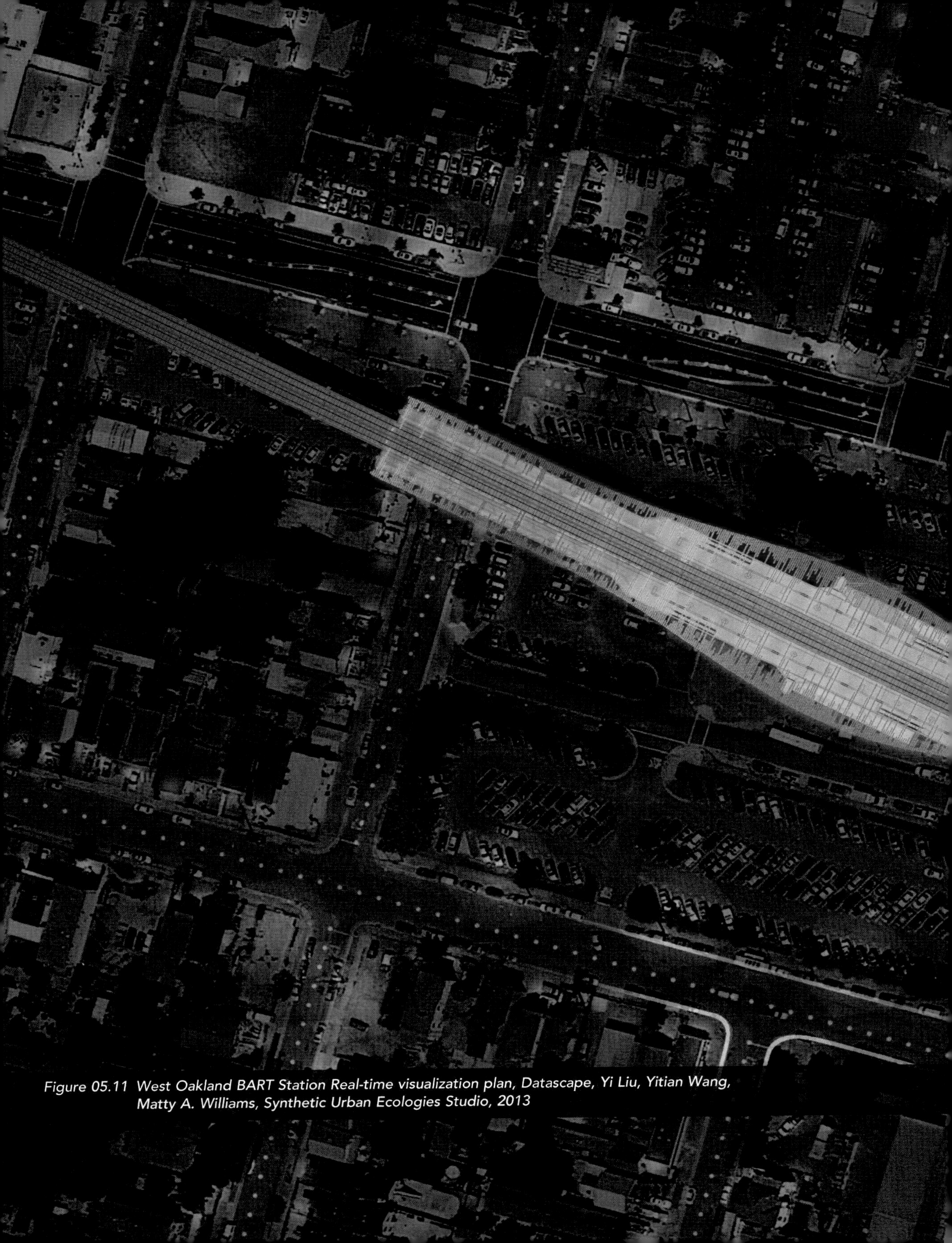

Figure 05.11 West Oakland BART Station Real-time visualization plan, Datascape, Yi Liu, Yitian Wang, Matty A. Williams, Synthetic Urban Ecologies Studio, 2013

atascape
akland, California, 2013

Yi Liu, Yitian Wang, Matty A. Williams, Synthetic Urban Ecologies Studio, LSU Fall 2013

Scope: Proposal for the City of West Oakland, California

Technology: Arduino, Firefly, Grasshopper

Datascape is a multi-component framework composed of a sensing system, data platform, visualization installation, and cross-communication platform choreographed in real-time to elucidate air quality issues in West Oakland. The framework is implemented through a website with live mappings of current conditions, a mobile phone application that uses Google Maps to provide a less polluted and healthier route for pedestrians, and a public installation as a visually responsive skin to the West Oakland BART Station. *Datascape* "exposes hidden information" to fight for environmental justice and address air quality issues. Building from a master plan for the Port of Oakland, "Adapt Oakland" and developed by a local design non-profit, Urban Biofilter, the proposed framework enables the democratization of air quality information and instigates informed decision making with multiple scales of impact. This methodology displaces the localized phenomena of particulate matter through virtualization and builds a platform for visualization, management, and community action.

The proposal contains five major components: a sensing system, data platform, communication platform, visualization schema, and infrastructural implementation. The project was developed in response to the Synthetic Urban Ecologies Studio taught by the authors in the fall of 2013. The students were prompted to create an adaptive management strategy:

> Understanding the role of management within systems of adaptability and change defines the designer's role as a curator or manipulator of processes. How do responsive technologies play a central role in the monitoring and management of these systems? How does computational iteration (micro-timescales) and autonomy advance concepts of adaptive management?[24]

The sensing network logs environmental data from the Bay Area Air Quality Management District (BAAQMD) and particulate matter sensors that are located throughout West Oakland. This network records pollution concentrations of PM2.5, PM10, DPM (fine particulate matter 2.5 and 10, and diesel particulate matter, respectively), and Ozone in real-time and links the data to geospatial co-ordinates.

Weather attributes including temperature, humidity, wind, and photochemical smog are recorded through similar data points with each sensing array recording within a 250′ radius. This network provides a robust view of the atmospheric qualities throughout West Oakland.

Figure 05.12
Map interface, Datascape, *Yi Liu, Yitian Wang, Matty A. Williams, Synthetic Urban Ecologies Studio, 2013*

The data platform provides content that drives the visualization and implementation of the sensed data. Similar to Xively, a customizable data hosting site, an open-source data hosting platform allows crowd-sourced data from the sensing system to compile the data at different timescales. The visualization of this data makes it possible to make temporal comparisons developing an understanding of fluctuating air quality. The project focuses heavily on the democratization of data within the urban environment. The team approached this through the articulation of policies that govern data acquisition and access as well as through specific implementations that focus on the visualization of displaced data. The proposal was inspired by the use of radiation sensors by Japanese citizens near the Fukushima nuclear power plant disaster.

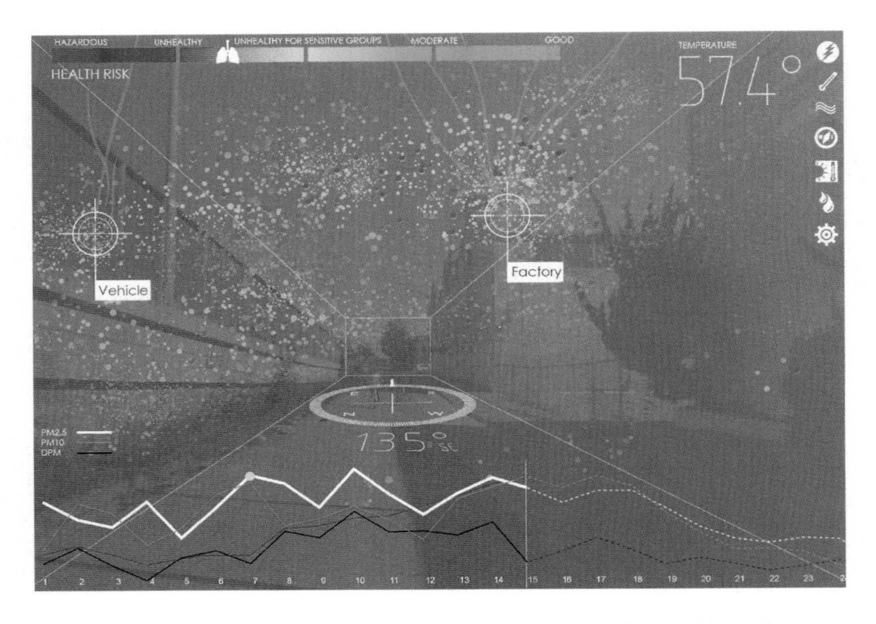

Figure 05.13 Perspective interface, Datascape, *Yi Liu, Yitian Wang, Matty A. Williams, Synthetic Urban Ecologies Studio, 2013*

The team developed a proposal for a framework that makes sensed data within West Oakland available to all citizens in real-time. Individual access to real-time data helps to shift the balance of power that has typically been in the hands of large multi-national corporations or government agencies. Conceptually this open framework would allow citizens, government agencies, and businesses to create healthier and more democratic decisions for West Oakland.

The first of the visualization methods is the mapping interface, which provides a contextual overview of the particulate matter levels in West Oakland. The interface provides the ability to toggle between individual sensor readings and to identify nearby emission sources. The mapping also provides a real-time health risk assessment as a composite overlay allowing users to make decisions regarding current dangers within the neighborhood. The composite map can display weather data such as temperature, humidity, and wind vectors as well as particulate matter and overall risk assessments. Beyond the real-time data that is located spatially, the line chart at the bottom displays historic, current, and projective data regarding overall pollution concentration.

Similar to the mapping interface, the second interface delivers an augmented view that overlays atmospheric data through users' smart

Figure 05.14
View of West Oakland BART Station from street, Datascape, Yi Liu, Yitian Wang, Matty A. Williams, Synthetic Urban Ecologies Studio, 2013

phones. The primary visualization is a color intensity that is overlaid on the interface as well as a health assessment that communicates the current environmental risk. The interface also highlights emission sources and indicates overall particulate matter levels through animated particles. Although the interface displays overall health risks and emissions sources explicitly, other information such as climate data is more subtle and requires users to have more prolonged exposure to understand the intricate relationship between each element.

In addition to the mapping and augmented reality visualizations the team also proposed a plugin for Google Maps that integrates route choosing with real-time sensor data. With sensor arrays at each

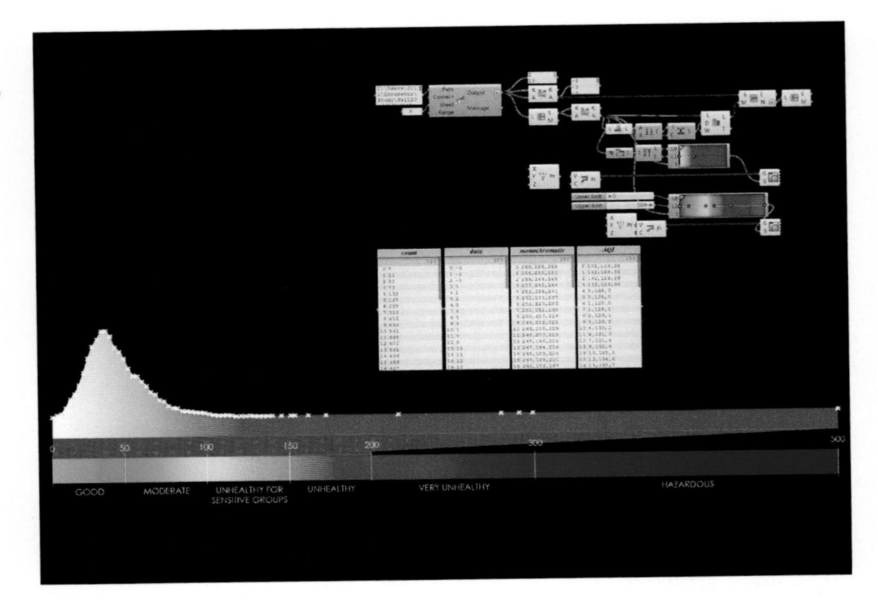

Figure 05.15
Grasshopper script for color range, Datascape, Yi Liu, Yitian Wang, Matty A. Williams, Synthetic Urban Ecologies Studio, 2013

intersection in West Oakland, the plugin uses the Google Maps API to propose the healthiest routes to users based on current risk assessments. Rather than using only time as a metric the user would be given the choice to take routes that have varying levels of risk based on current particulate matter levels at specific locations.

The proposal also imagines a communication interface that gives residents a private web forum to discuss monitoring efforts, mitigation, and current initiatives. This online space would visualize the current mood of neighborhood through the monitoring of hashtags, social media, and pollution levels to provide an overall, synthesized visualization of the community's current social state.

In addition to the sensing network and software based interventions, the proposal also speculates how infrastructure, architecture, and landscape might evolve to take advantage of this real-time data. The first of these speculations is at the West Oakland BART station where a lighting installation illustrates regional and local particulate matter levels. As a space, the station forms a confluence of user experiences that relate to specific times of day as individuals use the subway and visit businesses located within the area. The installation uses strips of LED lights to represent a one-minute change in regional particulate matter data and is related to local particulate matter data with an overlay of color using a monochromatic gradient. The visualization emanates from the center of the station and continues across the plaza adding a new reading each minute. The team also

Figure 05.16
Phasing map, Datascape, *Yi Liu, Yitian Wang, Matty A. Williams, Synthetic Urban Ecologies Studio, 2013*

imagined how street lighting might be utilized to integrate visualizations that would quickly visualize real-time particulate matter data. This system would be integrated into street lights, embedded within curbs, and within raised pavement markers along major diesel truck routes. As a system, this lighting overlay integrates with the installation at the BART station to provide an ambient display of current health risks.

As the system is currently imagined, it addresses particulate matter pollution, but would be flexible enough to incorporate an evolving range of community needs. At the same time, the open data platform and visualization methods give residents new tools to confront government and corporate power. As the community grows and pollution decreases, the sensing network and visualization methods will change accordingly. The proposal represents a comprehensive approach to how sensor systems, monitoring, and visualization can play an active role when in the hands of community groups and individuals. As a platform, the neighborhood becomes a canvas for methods of visualization hosted within the city and through software applications.

NOTES

1 Alan Berger, *Reclaiming the American West* (New York: Princeton Architectural Press, 2002), 151.

2 John Brinkerhoff Jackson, "The Vernacular Landscape," in *Landscape Meanings and Values*, Eds. David Lowenthal and Edmund C. Penning-Rowsell (London: Allen and Unwin, 1986), 65–77. Cited by Berger, *Reclaiming the American West*, 151.

3 James Carey, "Technology and ideology: the case of the telegraph," *The New Media Theory Reader*, Eds. Robert Hassan and Julian Thomas (New York: Open University Press, 2006), 243. Previously printed in *Communication as Culture* (London: Routledge, 1989).

4 Ibid., 242.

5 Andrew Blum, *Tubes: A Journey to the Center of the Internet* (New York, Ecco, 2012).

6 Kazys Varnelis, Ed., "Invisible City: Telecommunication," in *The Infrastructural City: Networked Ecologies in Los Angeles* (Barcelona: Actar, 2009), 122.

7 James Corner, "The Agency of Mapping: Speculation, Critique and Invention," in *Mappings*, Ed. Denis Cosgrove (London: Reaktion Books, 1999), 226.

8 Bennett, *Vibrant Matter* (see chap. 1, n. 26).

9 Weintraub, *To Life!*, 7 (see chap. 1, n. 21).

10 Seth Denizen's lecture "Immanent Histories" from the Symposium *Architecture in the Anthropocene: Encounters Among Design, Deep Time, Science and Philosophy*, curated by Etienne Turpin, 2012. Cited by Elizabeth Ellsworth and Jamie Kruse, *Making the Geologic Now: Responses to Material Conditions of Contemporary Life*, (Brooklyn, NY: Punctum Books, 2013.

11 Chris Reed and Nina-Marie Lister, "Parallel Genealogies" in *Projective Ecologies* (Cambridge, MA: Harvard University Graduate School of Design, 2013), 26.

12 Ibid.

13. United States Census Bureau, "About the Bureau," http://census.gov/about/what.html.

14 Aerial Photography Field Office, "Imagery Programs: National Agriculture Imagery Program," *United States Department of Agriculture*, http://apfo.usda.gov/FSA/apfoapp?area=home&subject=prog&topic=nai, June 10, 2014.

15 Harold Edergton, *Electronic Flash, Strobe* (New York: McGraw-Hill, 1970).

16 Heejin Lee and Jonathan Lievenau, "Time and the Internet," in *The New Media Theory Reader*, Eds. Robert Hassan and Julian Thomas, (New York: Open University Press. 2006), 266. Previously published as "Time and the Internet," in *Time and Society* 9, no. 1 (2001): 48–55.

17 Ibid., 267.

18 Brian Davis, "Anti-Terraforming and Ecosynthesis, Planetary or Otherwise," *faslanyc* (blog), June 19, 2012, http://faslanyc.blogspot.com/2012/06/anti-terraforming-and-ecosynthesis.html

19 Lydia Millet, "The Child's Menagerie," Opinion Editorial, *New York Times*, December 9, 2012. New York edition.

20 Nataly Gattegno and Jason Kelly Johnson, "Theater of Lost Species," Future Cities Lab Website, Accessed March 01, 2015, http://future-cities-lab.net/theater-of-lost-species/.

21 Ibid.

22 "Architecture as a Technicolor Canvas | 'Emergence' by Obscura," *Vice Creator's Project*, 2013. Film. https://youtube.com/watch?v=N2ZMH1XIxBI

23 Ibid.

24 Bradley Cantrell and Justine Holzman, "Synthetic Urban Ecologies project statement," Reactscape (blog). http://reactscape.visual-logic.com/teaching/synthetic-urban-ecologies/project-2-0-strategy-and-implementation/. This course was taught at the Robert Reich School of Landscape Architecture, Fall 2013.

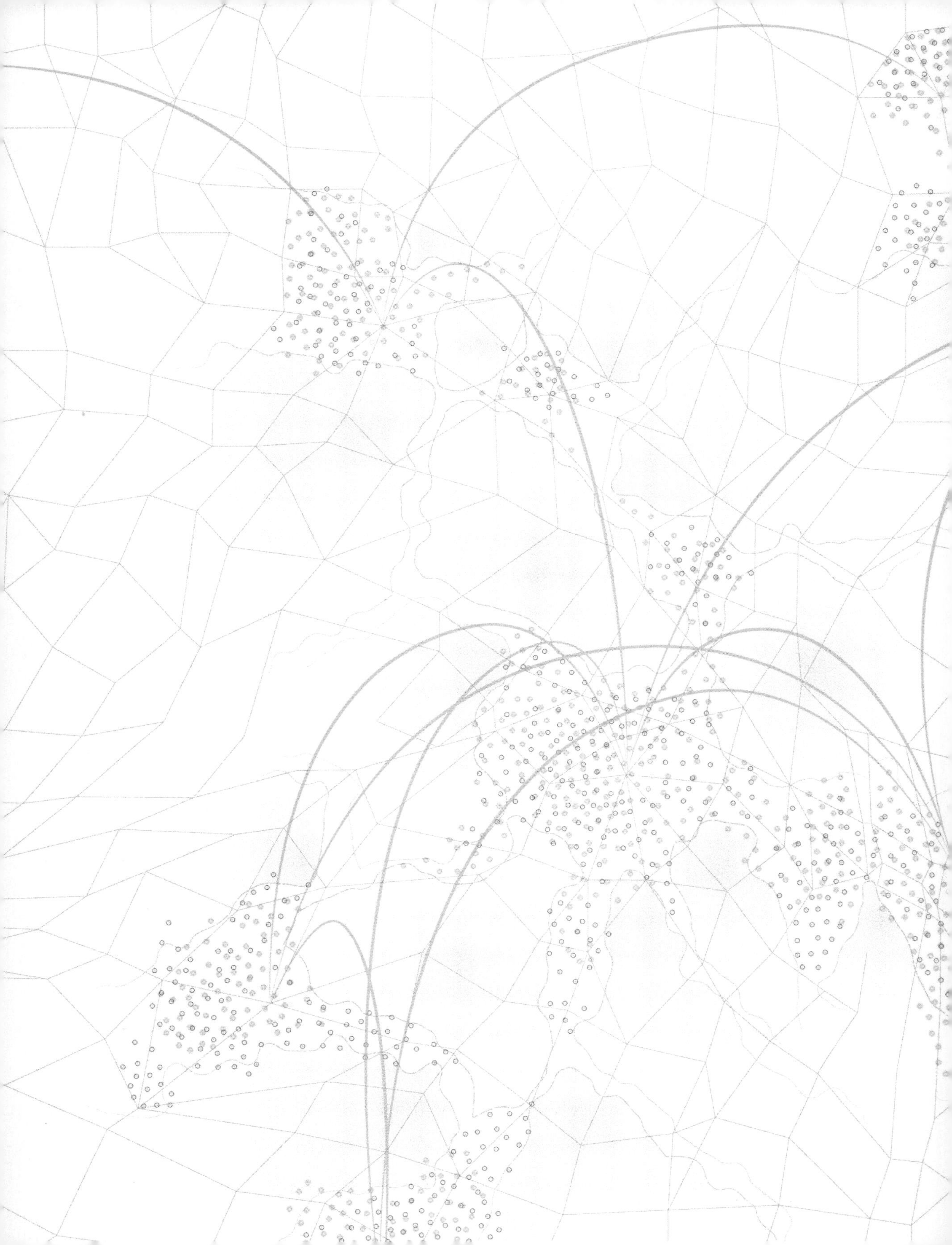

DISPLACE

06

Displace describes a removal and reconfiguration of information through physical and temporal displacement. Through the displacement and alignment of information, the connection between disparate moments is heightened. The process of displacement centers around context and our perception of phenomena based on contextual cues. The re-contextualizing of phenomena or subject requires a recalibration of the known to understand new conditions. Displacement may bring the utilization of a device or intervention into a new context as a way to understand new relationships within a greater field. It may also target a single phenomenon and help to understand its complexities and inner workings through new relationships. Displacement is fundamentally about removing landscape barriers—for the movement of materials, transportation, passage of information, and the ability to "see" or sense the landscape from a distant location. The physical separation of action from context is critical to this mode of response.

The origin of *Displace* is tied to the technological invention of the telegraph, marking the moment when information and communication were separated from transportation, "The telegraph not only allowed messages to be separated from the physical movement of objects; it also allowed communication to control physical processes actively."[1] The transfer of messages and ideas were no longer bound by physical limitations—decisions and actions, potentially with material and landscape consequences, could be displaced from spatial contexts. The technology and processes associated with the telegraph "provided an analogue model of the railroad and a digital model of language."[2] As the institution of rail across the United States formed, this early model of digital language through telegraph communications altered the possibilities for orchestrating off-site actions. "It coordinated and controlled activity in space, often behind the backs of those subject to it."[3]

Methods of displacement may be slight modifications or radical shifts. Information may be resequenced within a process through temporal displacement, closely related to strategies within compress. Methods, in addition to or divorced from temporal displacement, may be exposed to a physical realignment. The telegraph, rail, the automobile, and subsequent technological advances increasing mobility, transportation, and communication in relation to temporal co-ordination all "address the problem of 'distance'" in rural and urban contexts.[4] In Nicola Green's article "On the Move: Technology, mobility, and the mediation of social time and space," she describes the impact of "changes in physical proximity and distance" primarily altered by technologies associated with time and distance. She is particularly interested in how these advances have shaped the consciousness of temporality in the twentieth century:

> The central argument is that throughout the 20th century, changes in physical proximity and distance—including the effects of technologies designed to address time and distance, as well as a shifting consciousness of temporality—have 'dislocated,' 'disembedded,' and 'disembodied' individuals from local, collective, and co-present understandings of, and activities in, time, by 'stretching' social relations.[5]

One such transformative technology is the airplane. Denis Cosgrove describes the effects of the airplane, which—more than any other modern technology in the twentieth century—has "annihilated space by time."[6] The ease and "frictionless" movement provided by flight removed the barriers associated with travel by land and sea, "The boundaries that . . . fragment terrestrial space disappear in flight, so that space is reduced to a network of points, intersecting lines and altitudinal planes."[7] The aerial and bird's eye view would augment perceptions of the landscape forever. In 1959, NASA successfully launched the first satellite into space—the Explorer 6, a small, spheroidal satellite that transmitted the first images of Earth from space. In 1972, the first civilian Earth observation satellite, Landsat 1, entered Earth's orbit. This was the same year the United States successfully sent a man to the Moon.

> The ability to see the world from above culminated in the Apollo photograph of the spherical Earth, whole and unshadowed, taken in 1972. Responses to that image have stressed the unity of the global vision coupled with the need for local sensitivity in a globalized world.[8]

It is in the wake of this monumental feat of distance—the travel of man outside Earth's atmosphere to the Moon—that the combination of increasing accessibility to modes of rapid transportation, aerial imagery exploration, and communication have made "the far seem near and the shocking merely normal, local cultures have become fully networked around the world." The resulting dissociation of local communities "from spatial stability towards shifting, temporal coordination" marks the turn of the century as a networked society.[9]

The idea of *displace* is also related to the physical displacement of biophysical materials. Materials naturally move and shift across the landscape, carried by dynamic systems and entering into cycles of exchange and transformation. Ecosystems, as an assemblage of materials, are constantly in flux. As material fluctuations take place and traverse the landscape through dynamic flows, new ecological conditions emerge through processes such as the transplant mechanism of river systems. Concepts in systems ecology negate the presence of defining boundaries in the landscape, rather ecosystems are in a constant state of exchange and movement. The non-equilibrium paradigm in ecology recognizes that landscapes are susceptible to disturbances, sudden shifts, and bifurcations, only anticipated through probability. The transfer and displacement of biological materials with new technologies proliferates disturbances, resulting in rapid changes and repercussions across large systems as ecologies are quickly transformed.[10] The introduction of non-native invasive species displaced from their originating landscapes and the subsequent transformation of ecological systems are consequences of moving beyond landscape barriers.

The transfer of humans by air travel between ecosystems across the world with immunities to disease have implications for public health. The most recent outbreak of Ebola suddenly required the need for transportation and public health infrastructure to operate as a global system. The largest "dead pig scandal" emerged when the displacement of 16,000 dead pigs were found floating in the Huangpu River in March of 2013, contaminating the drinking water supply of Shanghai with carcinogenic toxins, "20 times above national safety levels."[11] Unlike a fish kill, there was no logical ecological reason for the appearance of thousands of dead pigs. In this case, the mass dumping of pigs was the result of a crackdown on the illegal black market sale of dead or sick pigs in tandem with a law requiring farmers to send animals that die of disease or natural causes to processing pits. To avoid paying fees, farmers opted for the cheapest solution.[12] The lack of on-site infrastructure for the safe disposal of dead pigs placed an overwhelming burden on the natural transportation system of the river, stemming from shifting regulations across regional operations without regard to material consequences.

The shifting and dissolution of boundaries occur through globalization as a result of access and transportation as well as infrastructure and the co-ordination of landscape logistics. As humanity continues to remove landscape barriers through infrastructural and technological interventions, the operational logics of these technologies become increasingly inseparable from the landscape. Gary Strang, in his article "Infrastructure as landscape" constructs a case for an "infrastructure of biological complexity," in which infrastructure is an integral player in the bifurcations and feedback of ecological systems: "the system is so complicated that is has begun to take on qualities of nature itself and, therefore, presents the same threat of random catastrophe that nature does."[13]

Displace offers the ability to shift upwards in scale, providing connections across disparate and contextually unconnected landscapes. Given the radical geologic changes the environment is undergoing due to the unprecedented scale of human settlement and activities, it cannot be ignored that, "we do not simply observe it as landscape or panorama, we inhabit the geologic."[14] Given the pressing consequences of anthropogenic landscape change, the luxury of conceptualizing the landscape as a passive backdrop to human activity is gone. Timothy Morton's article "Zero landscapes in the time of hyperobjects" describes the expansiveness and pervasion of hyperobjects.[15] His examples of global warming and nuclear waste in the form of carbon and plutonium-249 exceed human conceptions of space and time, transcending physical and temporal scales of interaction. The current manipulation of Earth is territorial in scale, manifesting itself across political boundaries and ecoregions. The "recursive alignment" of communication and transportation processes controlling exchanges of "material and energetic resources" within local ecologies, create conditions across disparate landscapes that are "structured by socially networked urban stakeholders, which extend 'tentacles' or organized channels beyond the city boundaries."[16] The "network dynamics" of these material and societal systems are composed of "many different temporal rhythms and spatial distances played out in different configurations and domains."[17]

In addition to the unrivaled urgency of these issues associated with globalization, they occur at a scale not traditionally considered by the landscape architect. This posture asserts that our current processes and methods sit between an extremely fast cycle of change and the geologic continuum. The effects of the Anthropocene require speculation that is geologic in a broad range of scales, temporal, physical, and cultural while responsive to the local. In many ways, this shift in scale disregards and detaches the information from the constraints of scale, leading to problematic interpretations. In John May's essay, "Logic of the managerial surface," he warns of landscape designs and management practices which operate as a complex assemblage of dissociated data, that when "concealed within a language of scientific ecstasy . . . electronic control has little concern for the concept of scale." The consequences and defining qualities of the managerial surface are the products of "geometrical infrastructures of Modernity," "organized specifically to compensate for the friction of distance."[18] The abstraction of landscape phenomena engenders landscapes of logistics, manifesting across the landscape as regional distribution facilities, such as Amazon Prime facilities or Walmart facilities. Through architectural investigations, Walmart has been recognized by Jesse LeCavalier for "its deployment of buildings as fungible components of distribution networks" defining a role for architecture within an operational landscape of infrastructural logistics.[19]

The origins of *displace* are tied to the practice of mapping and representing landscape through abstraction, as a critical body of theory for translating landscape as information with regard to location. As connections and relations are drawn between disparate locations, acts of mapping have established methods for measuring and defining landscape characteristics. Overlay, as a method to elucidate relations, correlations, and discontinuities, has a history within landscape architecture. Ian McHarg's method of the overlay map is, in a sense, the virtualizing of space through defining characteristics by placing them on the drawing board.[20] McHarg's method of shifting transparencies to identify constraints and opportunities is a precursor to more sophisticated applications of displaced landscape data. Advances in mapping technologies welcome this shift from a concern with boundaries and delineations to gradients and flows.

Through overlay, shifting contexts, or removal of phenomena, relations are extracted. Contextual methods, such as linking ecological patterns between similar climates, are a way to compare and evaluate across contexts. Removal aims to isolate and deconstruct phenomena through the privileging of specific phenomena. With ubiquitous and connected personal devices tracked by GPS, there is tremendous potential for displacing landscape information and running analysis in real-time. Etienne Turpin and Tomas Holderness have made significant progress utilizing the framework of Twitter, Inc. with support from the SMART Infrastructure Facility of the University of Wollongong and the Jakarta Emergency Management Agency to address widespread annual flooding in Jakarta, Indonesia, during the monsoon season. By harnessing social media to collect and display information, they are able to verify and translate this data into scalable mappings depicting flooding for Jakarta residences and governmental agencies. Individual moments and experiences coalesce into a cumulative mapping of tweets, presenting a real-time informational landscape tailored to the human experience of flooding.

Furthering the potential for the displacement of landscape information and contextual framing to shape our understanding and operations is simulation. Simulation as a form of displacement brings together multiple methods previously mentioned and moves beyond acts of mapping or representation. Simulation evolved from the realization that certain conditions can be synthesized to test scenarios without having to entirely re-create or re-build their contexts in the physical world. Simulation requires transferring information from a particular place to an interface, thus giving it a new reality or context. Contextual displacement through both digital and physical interfaces leads to the extrapolation and reconfiguration of phenomena into virtual, hybrid, and augmented realities. In his book, *Spacesuit: Fashioning Apollo* Nicholas de Monchaux thoroughly researches forms of technological innovation and design associated with the Apollo mission that have shaped the field of architecture. Not coincidentally, early forms of simulation were developed while preparing for the Apollo mission. It took the necessity of experiencing a context impossible to inhabit, the Moon, to invent methods for simulation: "the media of Apollo, and in particular its simulations, have transformed the nature of contemporary reality as much as the fact of a lunar landing."[21] De Monchaux also asserts that the "million or more semiconductors" used in the design of the simulation devices, directly influenced the "birth of the modern computing era," and ironically, "the very techniques designed so assiduously to protect and help transport the physical bodies of Apollo astronauts through the universe have directly contributed to the disembodiment of contemporary digital life."[22]

The technological leaps across distances of space and time have left a resounding impact on the landscape. In connecting this condition back to the non-equilibrium paradigm, Kristina Hill contends, that as designers "grapple with the disappearance of fixed boundaries," increasing value is placed on the "use of temporary spatial features, and the critical significance of processes." *Displace*, as a mode of operation fits well with her suggestion that "boundaries should not be treated as real biophysical phenomena, but rather, be stretched, shrunken, and re-envisioned across multiple landscape scales" to welcome the dynamic qualities of the landscape, "not as artifacts that deserve permanent memorialization simply because they once existed."[23] The projects outlined in this chapter speak to possibilities for addressing the *displaced* from the collective and present. Through strategies of re-locating, re-embedding, and re-embodying, there are potentials to design contextually significant relations through the application of responsive technologies.

Figure 06.02 Living Light, *The Living, Seoul, Korea, 2009*

Living Light

Seoul, South Korea, 2009

The Living

Collaborators: David Benjamin and Soo-In Yang

Scope: 20′ × 20′ × 15′ Pavilion

Technology: LED Lighting Visualization Driven by Air Quality Index Data

---

Living Light is a pavilion located in Seoul, Korea, designed to visualize air quality through a real-time light map. Using data from neighborhood air quality monitoring *Living Light* creates a connection between specific neighborhoods and the invisible and ephemeral phenomena of pollution. The pavilion creates a tangible and occupiable space that projects from the local scale to the city scale; this relationship displaces environmental data that is territorial and makes it comprehensible as an abstracted visualization. The pavilion orients and contextualizes users to the city, similar to a street map or diagram of public transit, through an abstraction or diagram. As a prototype, *Living Light* suggests a form of communication between architectural objects, users, and environmental phenomena. This communication creates a canvas that can begin to register phenomena that is ephemeral and difficult to comprehend, giving the data legibility through displacement.

Located near the Seoul World Cup Stadium, a busy urban area, *Living Light* is a 20′ × 20′ × 15′ structure, the pavilion is comprised of 27 panels—each representing a neighborhood boundary in Seoul. The panels are then bent to form the overhead canopy of the pavilion. The pavilion is lit through a series of LEDs embedded into a mapping that corresponds to the air quality in the associated neighborhoods. The visualization is curated to display several modes. To visualize air quality improvement over time, neighborhoods where the air quality has improved from the previous year are lit up while areas showing no improvement remain dim. To show real-time air quality, the LEDs shut off each hour and then light up in the order of best to worst air quality measurements. Collective interest is also expressed by directly communicating with the *Living Light* hotline by sending text messages with a neighborhood zip code. The hotline will respond with current air quality and the panel will blink to show interest. As individuals are interested in a neighborhood's air quality the panel will blink more rapidly.

The project examines regional visibility as a method of displacement that inspects a single thread of a much more complex system. As stated by Therese Tierney and Anthony Burke, "Infrastructural and

Figure 06.03
Living Light, *The Living*, *Seoul, Korea, 2009*

organizational networks today are complex, yet tend toward a natural state of invisibility."[24] Rendering phenomenal systems in this manner is a method of abstraction, diagramming evolving data streams into accessible forms. Dissimilar to design workflows, these tangible installations at one moment attempt to bring clarity and at another moment simply strive to raise interest in the subject. It is often enough to ask participants to dig deeper, to create cognitive relationships that are not possible without the displacement and subsequent abstraction of the environmental data.

Conceptually, *Living Light* speaks to a form of infrastructure that is directly connected to issues of public interest. As a prototype, the project tests the displacement of environmental data and its retranslation at the local scale. The pavilion provides a minimal and trivial connection to larger environmental issues but expresses a line

Figure 06.04
Living Light, *The Living, Seoul, Korea, 2009*

of inquiry positing access to data that is typically difficult to discern, providing citizens with the power to respond. The ability for our landscapes, architecture, and infrastructure to express information abstractly through accessible visualizations and data streams allows a relationship with physical objects. This evolving communication creates the possibility to interact with physical systems in a more articulated manner.

Scent Garden

Xi'an China, 2011

Rodolphe el-Khoury

Collaborators: Drew Adams, James Dixon, and Fadi Masoud

Scope: Constructed Garden and Pavilion

Technology: Distilled Scents and Responsive Scent and LED Poles

---

Scent Garden is designed as an installation for the 2011 Horticultural Expo in Xi'an China to provide a multi-sensory experience centered around scents of vegetation across China. The project confronts the favoring of visual aesthetics by emphasizing scent as a driver of landscape experience. The garden presents itself as a public space for gatherings or personal contemplation. Alternatively, the designers present the space as a "fragrance boutique" that forms an "electronically mediated multi-sensory experience."[25] It is this manufacturing of experience that becomes interesting as it creates a displacement between environmental inputs of smell and typical visual cues. Troposmia and euosmia are sensorial distortions that make alternate connections between "natural" odors and the interpretation of the sense as it is processed neurologically.[26] *Scent Garden* explores a version of this displacement, dislocating spatial and visual cues with olfactory information.

The garden manipulates the relationship between olfactory senses and the visitor's visual interpretation of landscape or environment. Visually, the garden was intentionally designed with a muted visual palette, both formally and texturally, to create a softened backdrop for the olfactory experience—forcing the commonly dominating sense of sight to the periphery. Dark gravel walkways, steel benches, and grey stone further subdue the visual palette and support a framework for gathering and circulating throughout the space. The garden is defined through plantings of rosemary, thyme, citrus, and groves of conifer plantings to create simple connections to the landscapes of Xi'an. This palette heightens and exaggerates the olfactory experience and contrasts the architectural spectacles within the exposition.

Choreographing the sense of smell, the designers push the concept of displacement, creating an environment that alters our perception of how landscapes are composed. The scents in the garden are distributed through a series of motion sensor activated scent poles that diffuse a range of fragrances derived from plant extracts. The scent poles are stainless steel tubes that disperse the scent from integrated cartridges. The scent is distributed by air currents or expeditiously through a motion activated fan as individuals approach the poles. In addition, the poles also have integrated LED lights that

Figure 06.05 Bottled fragrances in the garden pavilion, Scent Garden, el-Khoury, Adams, Dixon, and Masoud, Xi'an China, 2011

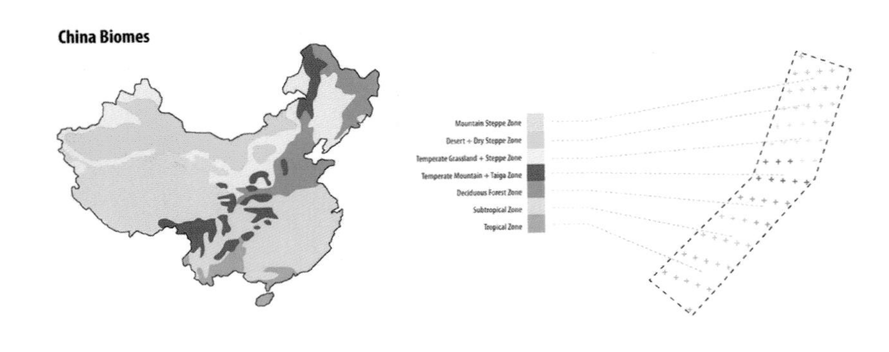

Figure 06.06
Scent Garden, *el-Khoury, Adams, Dixon, and Masoud, Xi'an China, 2011*

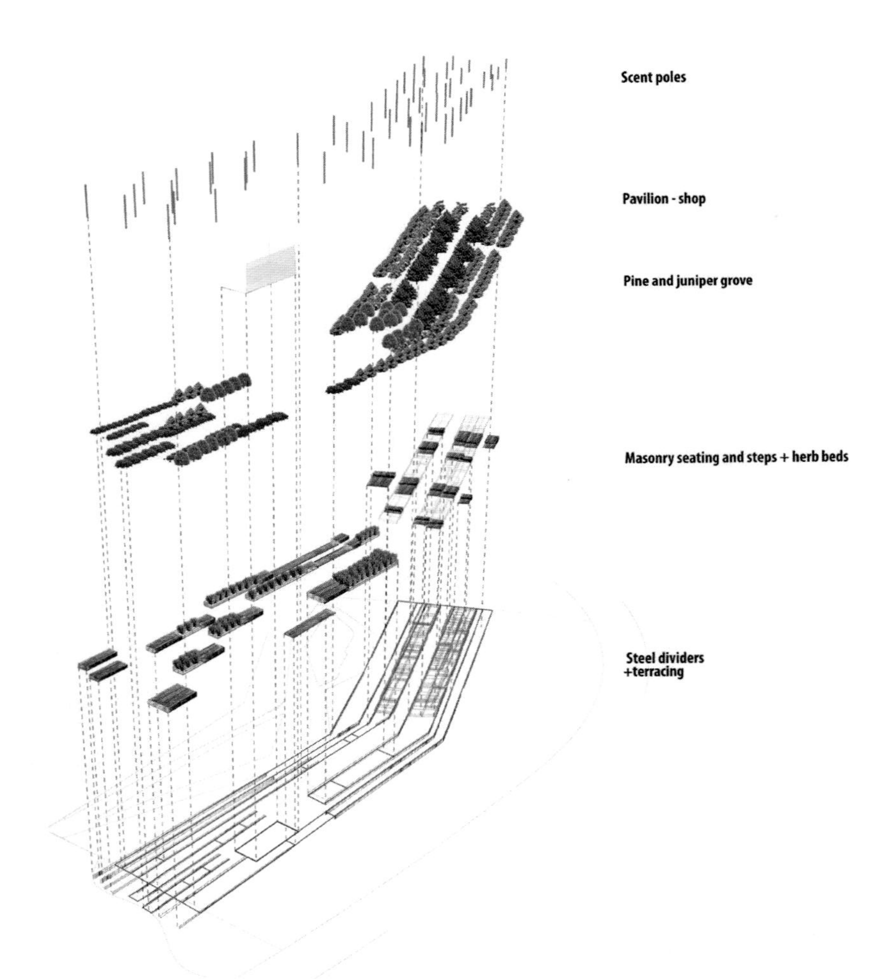

Figure 06.07
Scent Garden, *el-Khoury, Adams, Dixon, and Masoud, Xi'an China, 2011*

visualize the direction and intensity of wind elucidating the scent propagation patterns. The diffused scents dispersed throughout the garden represent specific regions of vegetation in China, forming an olfactory map. This displacement, the transfer of scent, from a larger region to the scale of the site produces a dislocated experience and begins to deconstruct normalized sensorial frameworks.

Further amplifying the manufactured experience, the garden pavilion houses the fragrances that are distributed within the garden for purchase. Bottling and packaging the experience further deconstructs the experience as the artifacts of the garden travel away from the site with the visitors. The act of removing the scent from the site packages the olfactory sensory experience as "a collectible," a manipulation of the typical stroll in the park. This can be seen as a form of distillation as the complex layers of scent in the environment are presented as single sensory experiences.

Figure 06.08
Scent Garden, *el-Khoury, Adams, Dixon, and Masoud, Xi'an China, 2011*

The *Scent Garden* provides an important venue to explore the deconstruction of sensory experience, choreographing new overlaps between the visual and the olfactory. This displacement highlights the composite nature of our distinct senses and alters our perception of landscape. The distillation and re-appropriation of scent provides a method to imagine ways designers may remix experience through new sensory relationships. Beyond the simple remixing for single installations, a similar method could be applied in the examination of novel ecological relationships, focusing on ecological fitness and sensory experience.

Datagrove

San Jose, California, 2012

Nataly Gattegno and Jason Kelly Johnson, Future Cities Lab

Collaborators: Ripon DeLeon with Osma Dossani and Jonathan Izen, assisted by David Spittler. Elliot Larson created the Twitter trends

Fabrication: MACHINIC Digital Fabrication & Consulting, San Francisco

Scope: Commissioned by ZERO1 (Jaime Austin, Curator), San Jose Public Art Program, National Endowment for the Arts (NEA) as a temporary installation

Technology: Text to Speech Module by TextSpeak, Arduino Mega and Uno, WiFly Shield by Sparkfun, Verizon Mifi, LCD panels by Sparkfun, LEDs by super brightleds.com (waterproof RGB strips), IR sensors by Sharp

---

Datagrove, designed by Future Cities Lab, is part of a larger group of projects and research attempting to design spatial interfaces to materialize the ubiquitous and almost always invisible infrastructure and associated phenomena of data that quietly traverses the landscape of the internet. *Datagrove* is an installation composed of LEDs, speakers, and LCD displays designed to render concealed data streams from multiple sources passing through the urban environment, whispering Twitter and other social media information within a protective and contemplative space. *Datagrove* temporarily resided in the courtyard of the historic California Theater as one of a number of distributed exhibitions throughout Silicon Valley and the greater Bay Area, for the ZERO1 "Seeking Silicon Valley" 2012 Biennial. Designers and artists were asked to illustrate the complex, mostly invisible, and "networked nature" of technological innovation in the particular region of Silicon Valley through the intersection of art and technology.

Designed as a "whispering wall," *Datagrove* challenges the personal and immaterial experience of social media across individual devices and asks how providing physical and spatial connections to this information might shape architecture to influence social, political, and environmental behavior in new ways. The looping structure of ribs holds glowing orbs that whisper Twitter feeds—indicating intensity by glowing softly or strongly—while offering "shelter and a place of calm to contemplate data streams from sources near and far."[27] As someone approaches the installation, proximity sensors actuate both visually and aurally, engaging the viewer in this new public discourse as a dynamic extension of social media.

The responsivity of the grove was designed to render both environmental phenomena and digital streams of data drawing a spatial relationship between the two. *Datagrove* is constructed to sway and move with the wind, building on the metaphor of collecting "the data"

Figure 06.09 Datagrove, *Future Cities Lab, California Theater, San Jose, California, 2012. Photograph by Peter Prato*

passing through the atmosphere. The installation attempts to portray the aggregate or seemingly random messages of the city, almost as if the armature were capturing and extracting the data from the Hertzian landscape. Building off of Nataly Gattegno and Jason Kelly Johnson's intention to explore novel and experimental approaches to architecture at the intersection of advanced technologies, social media, and the internet of things, "Datagrove forecasts a world in which networked information is interwoven into the basic elements of the city—its bricks, mortar, building technologies and appliances."[28] The materiality and spatial accessibility of *Datagrove* speaks to this future vision of responsive and information laden environments.

In considering the qualities outlined in the chapter "Displace", the grove captures tweets *trending* within San Jose and Silicon Valley though not necessarily *originating* from or the subject of the local region. The information may be the subject of anything and displaced from an entirely different location. The siting of the installation draws upon this displacement to call attention to the context in which these streams of data are aggregating through shared discourse. *Datagrove* is unique, drawing from individualized Twitter feeds experienced through personal devices, in that it "amplifies this discourse into the public realm," instead of refreshing your personal

Figure 06.10
Datagrove, *Future Cities Lab, California Theater, San Jose, California, 2012. Photograph by Peter Prato*

and curated Twitter feeds.[29] The slight movement and the responsivity of the installation to proximity, gives the impression of "increasingly life-like characteristics" through which citizens are asked to "engage and participate in its evolution" in a way that brings public discourse back into public space to bridge the Hertzian landscape with the physical landscape.

Figure 06.11 Datagrove, *Future Cities Lab, California Theater, San Jose, California, 2012. Photograph by Peter Prato*

Pachube, Xively, and Thingful

Web-based Platform, London, UK, 2008

Usman Haque, Haque Design + Research, Umbrellium

Scope: Internet of Things

Technology: Web Hosting and Common Services

Pachube, then *Cosm*, and now *Xively* was initially designed by Usman Haque as a web-based service for users to store, share, and access real-time sensor data through feeds. The platform is designed to make real-time information accessible by connecting physical and virtual worlds through remote monitoring. As a platform, *Pachube* was an important step in environmental sensing that acknowledged the need for a common service that brokered sensor data and hosted feeds for public or private access. This open platform to stream real-time data is important for several reasons. The first is the ability for a range of constituencies to have access to information directly without it being processed or cleaned up by outside interests. Direct access to real-time feeds creates a conversation where multiple vantage points can interpret and understand the information. This forms new opportunities to develop interesting visualizations, applications, and hardware that take advantage of databases that are consistently updated in near real-time.

Pachube evolved through several iterations as the platform was converted into a commercial product after being purchased by Logmein in 2012. Later, it was rebranded as *Cosm* and then most recently to its current iteration, *Xively*. While the underlying technologies have been strengthened, the open browsing of feeds by location have been minimized. The platform creates a hosting service that devices can write data to—for example data from a weather station can be written to the service using a common API. In a similar manner this data can then be made accessible and accessed by other devices that can connect to the hosting platform.

Thingful, a new effort by Usman Haque and his firm Umbrellium, creates a search engine for internet connected devices. While *Thingful* is not a hosting or brokerage service like *Xively*, it comes from the same thinking that attempts to provide platforms that make internet-connected devices easily accessible. *Thingful* acts as an aggregator for sensor data, providing individuals with locations to access the data. The information is browsable through a mapping interface and can also be searched based on keywords. *Thingful* spatializes data at the world scale, visualizing device locations and fidelity or resolution across the landscape. This overview and searchability makes it possible to quickly develop cross contextual relationships and to have a broader view monitoring effects.

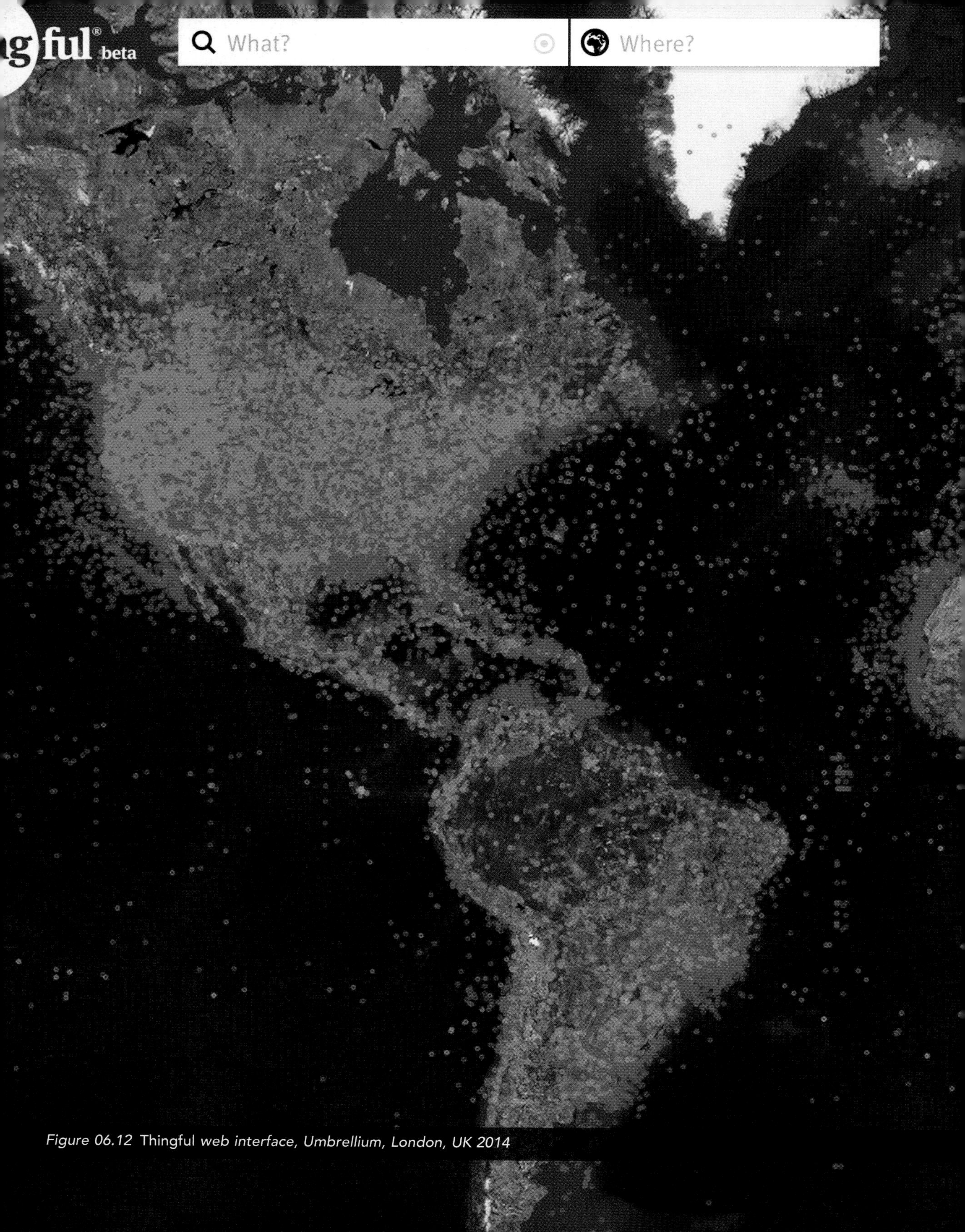

Figure 06.12 Thingful *web interface, Umbrellium, London, UK 2014*

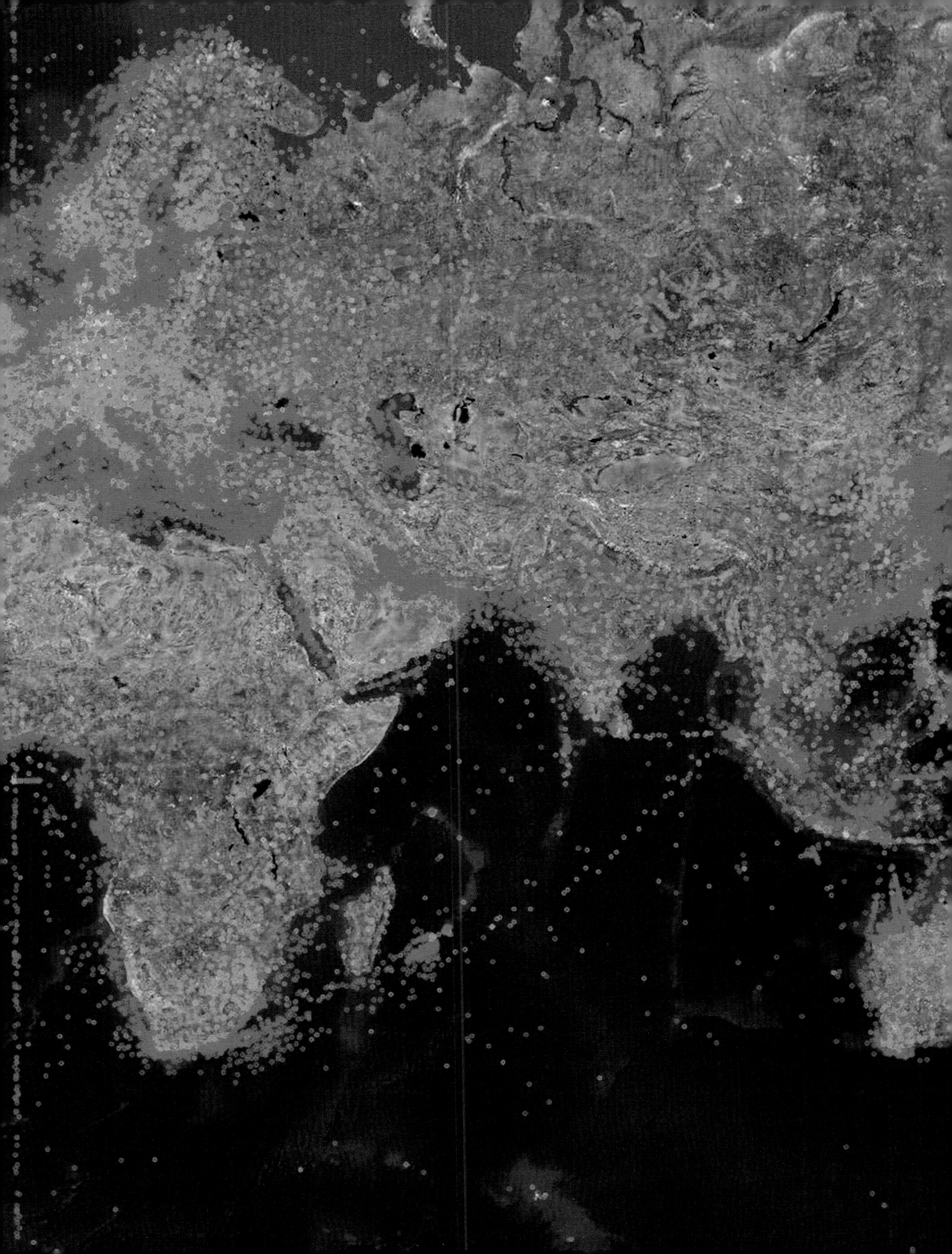

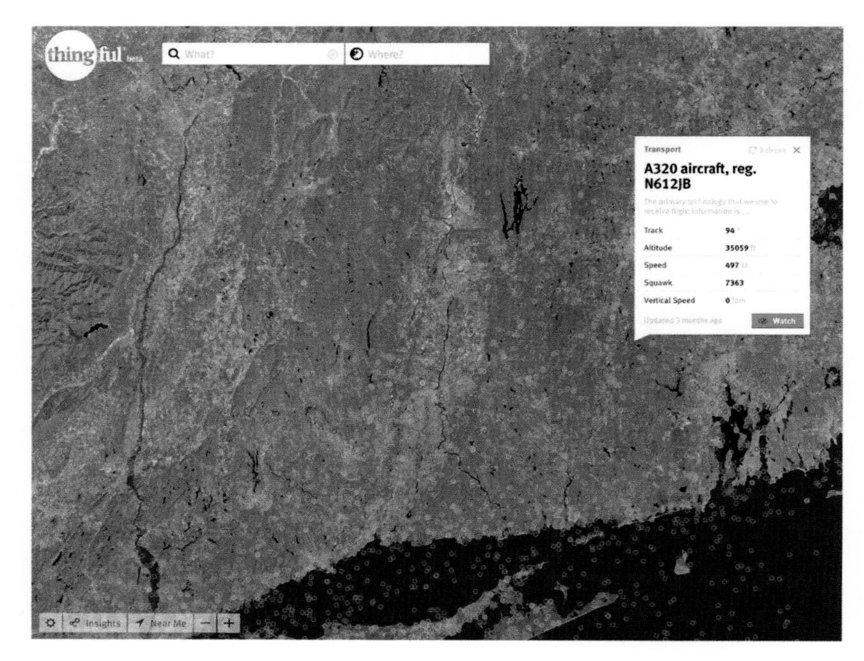

Figure 06.13 Thingful *web interface, Umbrellium, London, UK 2014*

Within design discourse, platforms such as *Pachube* are primarily acknowledged for their potential to shape architectural performance, particularly in the management of building systems, networked architecture, and connections to context or localized climate data. In *Sensing Space* authors Eidner and Franziska's description of *Pachube* hints that "In (the) future the web service should support architects and designers in the development of their buildings, by providing building-site-specific data in real-time, and beyond this also in the optimization of the eventual building use."[30] This form of connection with the environment is particularly interesting when we imagine the possibilities that form digitally connected ecologies. The platforms provide mechanisms to quickly prototype geospatial data and further both methods of displacement and new forms of connection.

Similarly, Landscape Architects have been working with geospatial data for the past three decades but platforms such as *Thingful* and *Pachube* posit a future where those data streams may be updating in real-time. This changes our paradigm of design as data flows into our software and we are able to traverse historic conditions and monitor ongoing changes, drafting, modeling, and illustrating proposals that continually update through linkages to data streaming in from the site. Hosting platforms such as *Pachube* provide the first steps to not only an "internet of things" but just as importantly models and simulations driven by continuous streams of site data.

Local Code: Real Estates

San Francisco, California, 2009

Nicholas de Monchaux

Collaborators: Natalia Echeverri, Liz Goodman, Benjamin Golder, Sha Hwang, Sara Jensen, David Lung, Shivang Patwa, Kimiko Ryokai, Thomas Pollman, Matthew Smith, Laurie Spitler

Scope: Multiple US cities have been used as case studies including San Francisco, Venice, Italy, Los Angeles, and New York

Technology: Python, ArcGIS, Grasshopper, Finches

Local Code is an analytical design tool produced by Nicolas de Monchaux for locating publicly owned underutilized sites within a city and producing landscape strategies specific to local conditions through combined geospatial analysis and parametric design. *Local Code* re-imagines the selected sites as a new urban system with optimized ecological performance at both the local and city scale. The concept for the project extends the work of Gordon Matta-Clark, particularly his project *Reality Properties: Fake Estates* 1971–1978, in which he identifies small plots of liminal spaces within Queens formed through anomalies and inconsistencies of various public property and zoning designations.[31] These lots, "demapped and operationally isolated fragments of New York real estate" and too small or strangely configured to build on, were sold at auction for twenty-five dollars each.[32] Matta-Clark purchased these properties and proceeded to collect as much documentation and information as possible from the city. Nicholas de Monchaux describes Matta-Clark's sites as both "virtual fragments as well as real."[33] In a similar manner, *Local Code* aims to identify remnant parcels through city data sources across larger urban landscapes to develop system wide relationships between seemingly untenable spaces. Through harnessing the local performative capabilities of individual sites, there is a collective unburdening of urban infrastructure and a coupling of ecological attributes, redefining the potential of physically disconnected sites as urban systems.

Local Code represents several projects by Nicholas de Monchaux that utilize a similar methodology for identifying parcels, mapping, analyzing, assigning design criteria, and running subsequent analysis related to costs and performance through parametric tools. His team designed a component—named "Finches" for Grasshopper—a visual programming plugin for parametric modeling in Rhinoceros 3D—building a connection between geospatial data and parametric software. The component Finches, allows for the import, batch processing, and export of geospatial data within Rhinoceros 3D. This process opens the potential for a continuous feedback loop between

Figure 06.14 Local Code Case Study Exhibit, Nicholas de Monchaux, SF MoMA, San Francisco, 2012

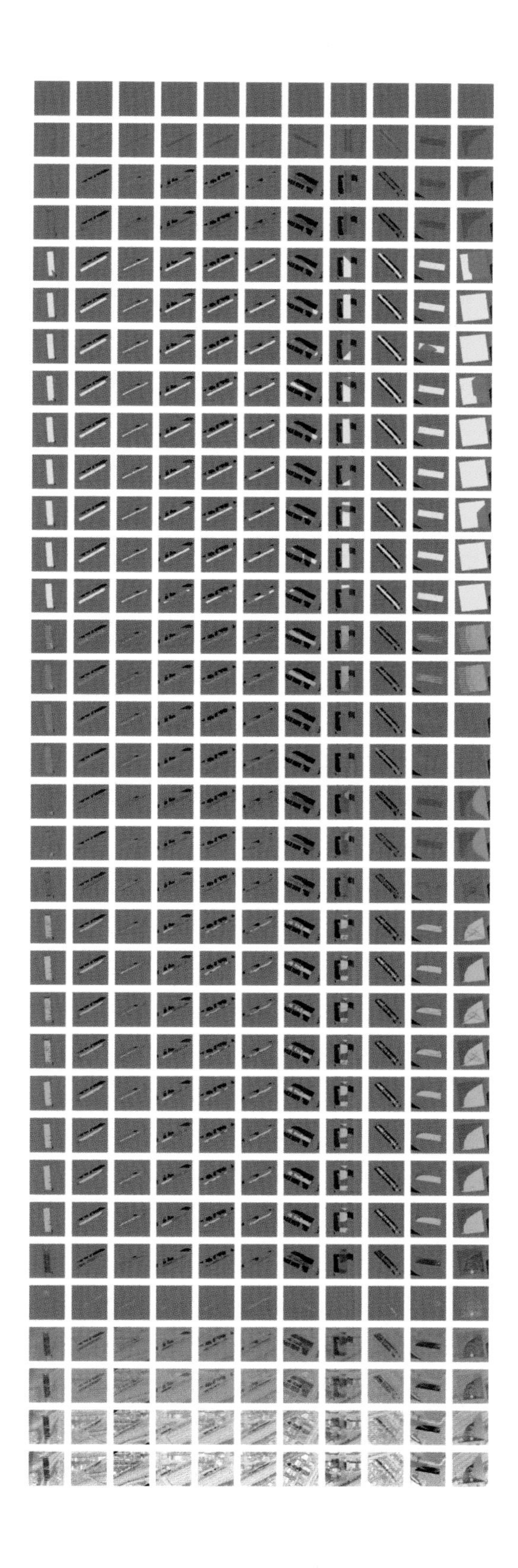

Figure 06.15
Process filmstrip, Local Code Case Study, Nicholas de Monchaux, San Francisco, 2009

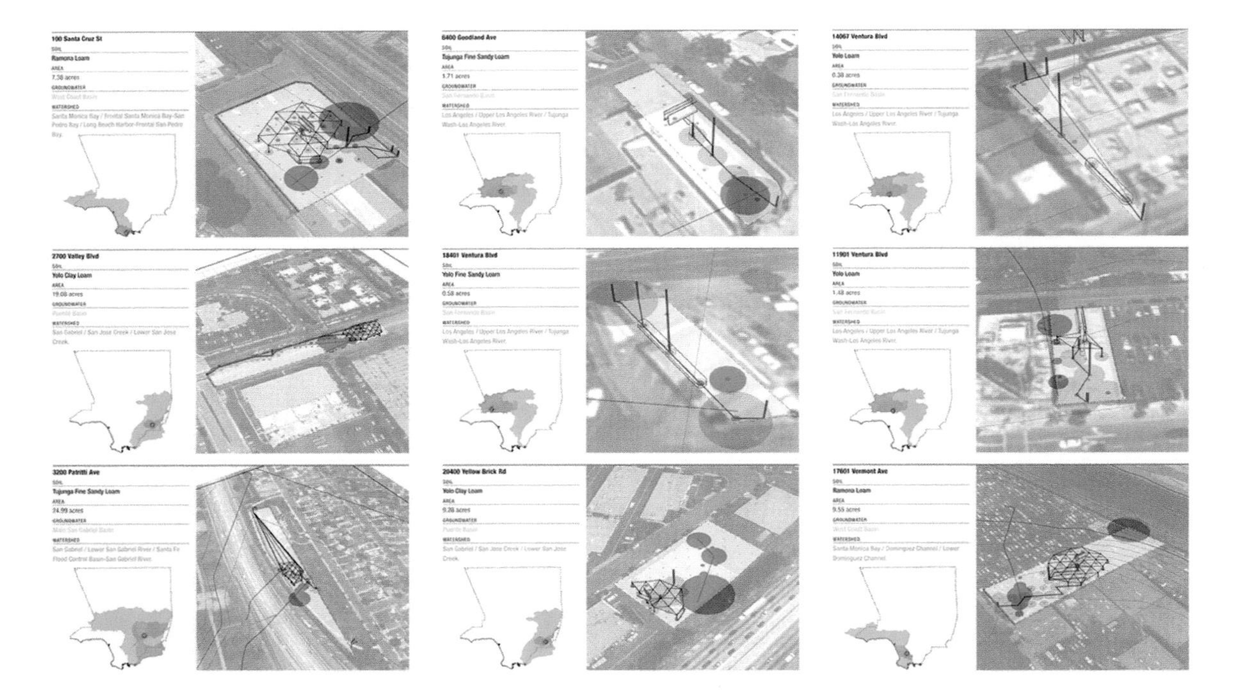

Figure 06.16
Local Code *Case Study, Nicholas de Monchaux, Los Angeles, 2010–2011*

design speculation and geo-referenced data describing site phenomena. The implementation of de Monchaux's workflow is used both to identify sites and to generate local design potentials linked to the larger urban network.

These potentials are visualized as design proposals, provocation, or performative templates. As a platform this methodology provides a dynamic feedback loop that expands the deterministic overtones that have stymied designers in their application of geospatial computational tools. The strength of this approach is the multiplicity of design iterations that all conform to a larger geospatial system connected by social, biological, and physical attributes. *Local Code*, as a practice of mapping, operates the way James Corner, in his seminal essay, describes how the agency of mapping "lies in neither reproduction nor imposition but rather in uncovering realities previously unseen or unimagined, even across seemingly exhausted grounds. Thus, mapping unfolds potential; it re-makes territory over and over again, each time with new and diverse consequences."[34]

The re-processing of site or system data to create virtualized networks that are "performatively" linked is the basis for understanding *Local Code* as a method of displacement. Data is aggregated through the design workflow and understood through the lens of parametric experimentation. This displaced environment becomes the fodder for experimentation, molded and manipulated through connections to performative data that drives the visualization design potentials, site decision, and form making possibilities. The data within the virtualized environment is explored through the abstraction of parametric design software and re-projected into the urban environment through design speculation. Design potentials are generated within the space of 3D modeling software, Rhinoceros 3D, virtually linking performance to site-conditions. This process creates a forum to discover new design strategies that operate both locally and system wide—and through the tuning and re-processing of geospatial data in relation to parametric modeling—the potentials are continuously rediscovered.

BILLBOARD PARK PROPOSALS

CASE STUDY

Figure 06.17
Local Code *Case Study, Nicholas de Monchaux, Los Angeles, 2010–2011*

Interview with Nicholas de Monchaux

Local Code

Nicholas de Monchaux is Associate Professor of Architecture and Urban Design at the University of California, Berkeley. He is the author of *Spacesuit: Fashioning Apollo* (MIT Press, 2011), an architectural and urban history of the Apollo Spacesuit, winner of the Eugene Emme award from the American Astronautical Society and shortlisted for the Art Book Prize; and *Local Code: 3,659 Proposals about Data, Design, and the Nature of Cities* (Princeton Architectural Press, 2015).[35] His architectural work has been exhibited at the 2010 Biennial of the Americas, the 2012 Venice Architecture Biennale, Chicago's MCA and San Francisco's SFMOMA. He is a Fellow of the MacDowell Colony and the American Academy in Rome.

01

Where did the concept for *Local Code* come from and how has the project evolved?

Local Code is a working method, deeply influenced by my own education in the late 1990s and early 2000s when the environment of architectural culture in both theory and practice was radically altered by digital practice, bringing enormous formal and expressive calculating possibilities with it. Yet, unlike the highly critically engaged practices it replaced, digital practice was mostly devoid of a meaningful theoretical and critical capacity. It was impossible to be a student at the time and not observe the way in which ideas were replaced by tools.

As a result, my initial inquiry, (which evolved into the book *Spacesuit: Fashioning Apollo*) was to figure out what a theoretically and historically engaged digital practice would be. And what methods or digital tools would compensate for what at the time were significant limitations of working within environments as complex as cities and landscapes. In my long detour into the history of space, and through that inquiry spending a lot of archival time within the closely related history of computing, it became abundantly clear to me that technology is a cultural artifact. Like any cultural artifact, it carries certain assumptions and those assumptions need to be challenged and questioned, particularly if the person who is using the software (the design practitioner) is different from the person who constructed the software (the software engineer). As one example, we are the recipients in architecture and landscape architecture of many hand-me-down tools from aerospace and engineering (parametric techniques as one example) that embed within them a culture and ideal of optimization. But even if you accept that an optimal wing can be designed, or an optimal technique for laying carbon fiber, it is equally true that a city or a landscape cannot be optimized, except across a very narrow range of variables. But unless an awareness of this cultural and institutional background accompanies an engagement with the tools in question, these sorts of assumptions are likely to slide by unobserved.

The *Local Code* project was part of a larger attempt to think about what might shape digital design practice that questioned, but also drew from, the actual cultural and material history of design technologies. In particular, to provide an alternative approach to the inherent tendency towards optimization, as we just discussed, which can be seen in the practice of parametric design as it broke upon architectural culture in the 2000s.

Having developed my own ideas about the real and messy qualities of cities and infrastructures (in particular using GIS in environments like Venice when I was teaching at UVA), I was deeply frustrated as a technologist and a user of technology, with the scripting and parametric tools that became widely available, and their inability to absorb or engage GIS data. In the case of parametric tools, we had this form of design media that was really good at dealing with information and designing according to given pieces of information, but we couldn't feed it any real or relevant information about the environments we wanted to design in the form of the spatial databases that are now so abundantly present in our encounters with cities and landscapes. When I sat down to start thinking about this kind of work in 2005, 2006, 2007, there were numerous ways to bring in GIS data as geometry into parametric software, but no method for accessing the associated database and attributes within a parametric environment—which of course is where all the interesting bits of data are.

As I was trying to develop these tools for dealing with GIS in a parametric environment, I was looking for datasets and data problems that might be conducive or interesting to try and hack this thing out. At the time, I was speaking to some friends in San Francisco, including an assistant to the then Mayor of San Francisco, Gavin Newsom (who was very interested in the use of technology by cities), and became aware of an existing database of 1,500 vacant city-owned parcels, cataloged but not maintained by the San Francisco DPW, called "Unaccepted Streets." With help from the Mayor's office, I pried this database loose from DPW (and in this instance 'database' was just a very grand term for an excel spreadsheet).

In the resulting set of explorations, I began to think about the process of exchanging GIS information with a parametric design tool to imagine tailoring specific landscapes for specific sites. Very much not with the megalomaniacal idea that one would actually go out and build all these landscapes! But instead to explore a new kind of ability and a new kind of visualization that could be used to advance a conversation about the value of infrastructure. In particular, I believe that the ability to visualize a thousand micro interventions is an important counterpoint to the grand visualizations of traditional mega-infrastructure threatening to control the discourse surrounding urban resilience.

02

Conceptually, *Local Code* has ties to the mapping work of Gordon Matta-Clark, but leverages computation as a mechanism to expand the scope and design potential of unconsidered spaces. How does this evolution from analog to digital method provide new opportunities and insights? Does it close any doors that may be important?

At the risk of being pedantic, I would say that one of the fundamental shifts of modernity is the recent rise of information in media; It has not even been a century since we were able to define "information," when Claude Shannon and others actually defined a signal or bit (this thing we now call content) and observed that the thing in the telephone wire is distinct from the telegraph wire itself. Yet, much in the same way that Colin Rowe and Fred Koetter took architectural culture to task in the 1970s for inadequately exploring the architectural and urban possibilities of collage, I believe that today's design culture is situated in an increasingly informational culture, but is very tentative in the way it has engaged the architectural possibilities of an information-aware practice.

For we are awash in information. Statistically, we now produce as much information every few days as we produced in all of human history until 2007. And none of which is necessarily what we want or need to figure things out. We have moved almost instantly in the current decade from a position of never having enough information about the places and situations that we seek to design for, to having too much information, but never quite the right kind—that's the particular nature of our current problem, and opportunity. Now, this shift is also simply a landmark along a much larger shift towards what one might call "an informational turn" that dates back to the first cadastral mapping of territory thousands of years ago and accelerated in the eighteenth and nineteenth century when we started to look at our environment as a place that could be described by information, and information separate from actual qualities of place; for example the development of thematic mapping.

In that larger arc of history, you could look at landmarks such as the invention of GIS in the 1960s or the invention of CAD as simply landmarks along an accelerating curve. These processes of design no longer require physically pushing graphite around or the McHargian method of layering physical Mylar, but now are firmly lodged in the realm of bits, yet without a very clear attitude towards that abstraction. The rhetoric of the "hand versus the computer" and associated pointless debates that plagued my own time in architecture school are reflective of the unconscious and poorly articulated anxieties that a shift towards a landscape of information brings. Now we see architecture move increasingly towards BIM and collections of databases, not documents, and the physical landscape ever more populated with sensors and data approached in novel ways. As a set of design professions, it is most unprofessional of us not to think deeply about what this shift towards information means.

I believe Matta-Clark is particularly interesting in this regard. The historical threads his life touches are particularly relevant: his father worked for Le Corbusier before abandoning architecture; Matta-Clark's first job was actually in the Binghampton urban redevelopment agency; and his architectural education at Cornell University took place during the height of an informationally-inflected formal analysis movement, whose intellectual leader was Colin Rowe—who arrived at Cornell in the same year as Matta-Clark. And of course, he [Matta-Clark] violently rejected this legacy, leading in its own way to the speculations on alteration, change, and the arbitrary nature of mapping that became known as *Fake Estates*—which is quintessentially an informational project, begun by tracking down newsprint catalogs of properties considered to be abandoned and auctioned off by the city, and by the search within those thousands of pages searching for the slivers or micro lots hidden within. Matta-Clark was deeply fascinated by these tiny triangles formed where two properties end. At the point of Matta-Clark's death, this collection of documentary material became an artwork, later assembled and exhibited for the first time in 1992.

To me, then, the project represents a tragic yet hopeful trajectory of thinking about vacant space that is unfinished but no less interesting for that. A year before his death, Matta-Clark received a Guggenheim Grant to create a community garden within a vacant lot, showing real engagement within the fabric of the city as a renovation in actual space rather than the collection of information about it. In some ways, he was returning to his roots in architecture to do something in which he was simply the design agent or instigator, and involved many other people and many other kinds of practice other than what Matta-Clark is famous for—which is this pure artistic selection of demolition or the creation of an aesthetic experience.

In the context of the *Local Code* project, then, Matta-Clark was definitely the inspiration and the catalyst. In thinking about this dataset and what the potential utility was, I was deeply influenced by

Matta-Clark's *Fake Estates* projects as an eloquent and unfinished essay on the utility and value of vacant lots in cities, particularly addressing how we think about urban territory and urban landscape. Building on this inquiry, my research on the use of vacant and underused parcels as green infrastructure led to the realization that this software bridge I was attempting to build could be crucial in many design problems.

It then took the next project, *Local Code: Los Angeles*, which was done in collaboration with a non-profit in Los Angeles, the Amigos de los Rios. This was an equally kind of imaginary project, looking at sites under billboards for their political and advocacy potential in LA, for this idea to serve a certain material purpose. And then the project in Venice, *Local Code Venice: Ecology of Strangers* (part of the 2012 US Pavilion in the Architecture Biennale) capitalized on the confines and dysfunctionality of Venetian politics to rethink abandoned pieces of territory in the Venetian Lagoon.

Going back to NY and bringing the project full circle, we are now collaborating with an urban ecology Lab at the New School lead by Timon McPhearson, who has mapped 30,000 sites to identify their potential for ecosystem services. We are also collaborating with a real group of non-profits, in particular a group called Green City Force focused on community urban greening with at-risk youth. I don't know what will come of this collaboration—it's a much more ambitious and long term agenda to engage large-scale visualization processes with an on-the-ground, small-scale intensive remaking of space. With some hard work, good luck, and good intentions, hopefully something more substantial and enduring will emerge.

03

***Local Code* has attempted to confront the limitations of McHargian mapping methods and common applications of GIS by applying rule-based constraints directly to design interventions. By capitalizing on the application of datasets, *Local Code* addresses critical regional issues within current frameworks, sidestepping traditional planning processes. How does this method change the role of mapping within the design process? In what ways does this method facilitate the designer's ability to confront indeterminacy?**

I have already mentioned the revolution of information and its impact on design practice and design culture. As a series of revolutions in medium, I think I would cast my timeline pretty far back to deeply fundamental inventions like trace paper, which didn't exist in design studios until the late nineteenth and early twentieth century or rapidograph pens. The computer is also an information tool, different in scale and capacity though no less revolutionary.

"Media," to be truly pedantic, comes from the Latin word for "lens" and is fundamentally a way of seeing. Even design media, which are ostensibly tools for creating things, are actually tools for seeing the world as well. In this regard, the text *Maps as a Mediated Way of Seeing* is a redundant title because "media" is inherently a way of "seeing." It's always struck me that as designers we are so caught up in the ways in which design media creates this imaginary reality we'd love to believe is real that we often don't understand the imaginary realities we draw and seek to bring into being are themselves methods of observation.

One of the things that is parallel and deeply influential to my own thinking is the narrative of the city as an emergent, self-organizing network, not in the kind of formally-focused, strident way that architects have come to talk about in the last 10 to 15 years as they seek to make software-inflected buildings, but rather the way Jane Jacobs describes it in her urban trilogy.[36] Particularly, in the last chapter of *The Death and Life of Great American Cities*, she talks about a single urban park and the impossibility of isolating all the variables contributing to its success and failures as a specific response to the contemporary attempts (at her time) to compute the city. It was that chapter and a parallel essay conceiving of "complexity" by Warren Weaver, her patron from the Rockefeller foundation, which led the Rockefeller foundation to contribute a substantial amount of money to complex systems research. This research, accompanied by numerous other events over the last 40 to 50 years, brings us here, in 2015, investing in understanding and predicting complexity, to then finding ways of acting loosely or non-predictively—ultimately forming a deeper understanding of the city itself as a network of networks. This concept of the city as complex and emergent includes both unexpected resilience and unexpected brittleness—operating simultaneously as non-isolate, systemic, and fragile. In this context, the ability to envision a piece of infrastructure existing on 1,500 small sites opposed to one big site, is an essential part of developing a suite of conceptual and literal tools for designing with the complexities of issues cities actually have instead of singular infrastructural interventions.

As we require cities to become resilient against climate change and colossal urban growth, the ability to design along the grain of a city's own complexity is extremely important. *Local Code* is, more than anything else, my attempt to indicate such a direction. From where I sit inside a School of Architecture and Landscape Architecture, I can focus on training students to conceive of tools and solutions that might be of better use than the tools available in praxis, where the luxury to think about tools is not always afforded.

04

Do you see core principles that exist within *Local Code* being adopted in other modeling, simulation, and mapping technologies?

I am definitely not the only person doing this kind of work. Some of the masterpieces of late twentieth century planning involve precisely this investigation of networks and connectivity in the city. The work of de Solà-Morales and others in Barcelona in the 1980s, for example, was deeply connective and focused on, for example, the singular quality of sidewalk surfaces—undramatic but transformative—long before the showcase pieces by which Barcelona became better known. There is also the work of Jaime Lerner focused on multiple small-scale interventions in Curitiba, also in the 1980s. This work was adapted and expanded by others including the transformative mayors of Bogotá in the 1990s—Antanas Mockus and Enrique Peñalosa, the latter focusing specifically on connectivity and networks of resources for all communities. In some ways, it is difficult to find good urban work that does not recognize in some way the fundamental idea that the city is a social, pedestrian, economic, and cultural network. Increasing the performance of that network to enable every person to get to every place when we need to in a reasonable amount of time, feel safe in public space, or move across the city and engage with others are fundamental qualities of urban transformation.

And in a way, I would contextualize still further; this thread in the history of planning goes back to Patrick Geddes' training as a biologist and his resulting perspective on cities as metabolic networks. *Local Code* is a small tile added to that mosaic in an attempt to better translate that thinking into the landscape of information networks that is increasingly dominating urban planning and administration without any of the lessons of the city as a network itself.

05

Can you speak to the role of designers as tool makers? In particular, our ability to extend existing software through access to SDKs, scripting, and visual programming.

If you're not using your own tool, you're using someone else's. You can be strategic about whose tool you use, but I've always been struck by the way really good designers are able to push tools as much as the content of those tools. And here I would include not just tools of representation (the revolutionary 1980s work by Libeskind, Koolhaas, and Hadid), but also tools of construction—like the contemporaneous integration of the architecture office and the building component manufacturer accomplished by offices such as Norman Foster's and Renzo Piano's. These breakthroughs are manifest in the resulting design work but dependent on the invention and application of tools with the overall workflow.

In this context it's interesting to observe that software engineering in the last 30 to 40 years has contained two conflicting impulses: between top-down and bottom-up design methods, identified by Eric S. Raymond in *The Cathedral and the Bazaar.*[37] McNeel's very success with Rhino, and part of why we use it for *Local Code*, comes from the way they've opened up their tools for hacking and innovation—notably with software like Grasshopper. Conversely, an environment like Revit is enormously powerful, but only in a particular way, and so encumbered it can't even meaningfully interoperate with Autodesk's other products. Architects and landscape architects have never been the most powerful actors in shaping the landscapes surrounding building and, in the same manner, we are not entirely in control of our media selves either.

It is very important, especially in places like schools of design, to introduce students to the notion of hacking and making software in the same way they are hacking and making their designs as a vital tonic to the fact that in most situations they will be handed a tool they can only shape or hack in the most elliptical ways. At the very least, the process of teaching and hacking tools in a utopian universe would lead to a set of modular software altered by the designer as much as the design—hopefully, leading to some level of both invention and understanding related to the effect of software that we can't change.

06

***Local Code* is a tool with very specific applications and outcomes. What are some of the interesting evolutions that have occurred as you and others have pushed the boundaries of *Local Code*'s application?**

The mind-blowing moment was not in the actual creation of the software (which was a lot like any design process; you have some sense of what you want and it's just a process of diminishing frustration until you get close enough) but sometimes with a sense of historical symmetry and echo. Indeed, the original idea for *Local Code* came very specifically from part of this translation from the world of technocracy to urban administration that I charted in the *Spacesuit* book. In particular, the work of Howard Finger, who moved from being NASA's director of nuclear propulsion research to being the first director of R&D for the newly created HUD. Under George Romney, HUD successfully prototyped an automated site-planning tool for prefabricated housing, but at the time it didn't go anywhere because of the paucity of digital information about site.

Yet another such discovery happened retrospectively and archivally as an intellectual discovery with Matta-Clark and the book I'm writing now about the *Local Code* work. I had no idea when I started this process that Matta-Clark was getting involved and interested in community development or that he was writing to digital labs at MIT and UCLA to figure out the utility of digital visualization. I wouldn't be so arrogant as to say I am picking up those threads in particular, but I am so curious as to what their evolution in Matta-Clark's own work might have been.

In the end, having spent a lot of time with the history of technology, I can only ever muster up a particular flavor of optimism. The one thing one can say about transformative technologies is that they transform in ways that are impossible to anticipate.

07

Are you excited about any current trends in data collection and mapping? Are there technologies or practices that you see as being influential in the near future?

We are in the middle of a revolution defined by the wide availability of satellite imagery of the globe. I am hopeful we will begin to get the wide availability of other kinds of information we have now in American Cities, essential for other types of transformations. Five years ago, I needed to get the help of the Mayor's office to track down and access DPW data. Today, data in SF is among the most open in the world, with anyone able to access such information from a web interface. Yet this kind of openness is the exception, not the rule.

In addition, I believe the coming widespread availability of other kinds of sensor data and the potential to mine these much larger informational systems for their anomalies—analogous to the way that *Local Code* does for the anomalies within a property database—will provide particularly important design opportunities. So I would come back to the fact that in regard to cities and landscapes, the big question is the openness and access of data. The one thing that has become abundantly clear to me is that the real landscape of the city —with all the contradictions and possibilities of shared space—is no longer precisely and abstractly mirrored by a virtual landscape of information, but is very much one in the same, a kind of inextricable mashup. Therefore, we need to be approaching questions of publicity, privacy, access just as strenuously in the world of information as we do in real public space. Not because it is analogous to the real space of the city, but because it increasingly is the real space of the city—just as much as the actual sidewalk.

Rapid Landscape Prototyping Machine

Owens Lake, California, 2011

University of Southern California's Landscape Morphologies Lab

Collaborators: Alexander Robinson, Andrew Atwood of Atwood-A and First Office is a partner in the project

Project Budget: Advanced Design Tools for Reconciling Scales of Use and Perception in Large-Scale Infrastructures initiated with a USC Arts, Humanities, and Social Sciences grant in 2011. Lauren Bon and the Metabolic Studio supported this research

Scope: This project undertakes designing and building a custom landscape prototyping machine in order to improve the design of dust mitigation landscapes at the Owens Lake in Lone Pine, California

Technology: Physical Sand Model manipulated with a 6-axis Robotic Arm equipped with a custom set of End of Arm Tools, Digital Projection, 3D Laser Scanner, and a Custom Software Suite

---

Alex Robinson and the Landscape Morphologies Lab have developed a sophisticated and computationally advanced design methodology for modeling and simulating future landscape processes and conditions in Owens Lake, California. The former Lake Owens, drained by the water-demanding Megalopolis of Los Angeles, is now the source of hazardous dust storms ironically requiring substantial amounts of water for mitigation. Through the hybridization of robotics, digital projection, and 3D scanning, Robinson explores the potential of autonomous land construction to mitigate hazardous dust storms by regrading the landscape to require less water and maintenance.

The *Rapid Landscape Prototyping Machine* (RLPM) is a method for virtualizing and projecting the physical world that pulls from methods of landscape construction, landscape modeling, and representation—then brings them together into a highly specific and multi-dimensional platform. Pulling from methods of both digital and physical modeling practices, the RLPM is composed of a 6-axis robotic arm with a custom set of end-of-arm tools designed specifically to engage with sand-forming processes such as digging, piling, pushing, and raking—thus, forming a uniquely landscape oriented method of digital fabrication. To test the potential of regrading areas within Owens Lake, a series of tooling paths are designed for the robotic arm without dictating a topographic form, rather the topography is a material response: "driven by algorithmic tool-paths, the medium simulates the sedimentary assembly of landscape."[38]

The use of digital fabrication (mainly CNC milling) as a method for designing topographic conditions in landscape architecture (as well as architecture) has been criticized for its superficial engagement of surface, ultimately filtering out the material and dynamic qualities of

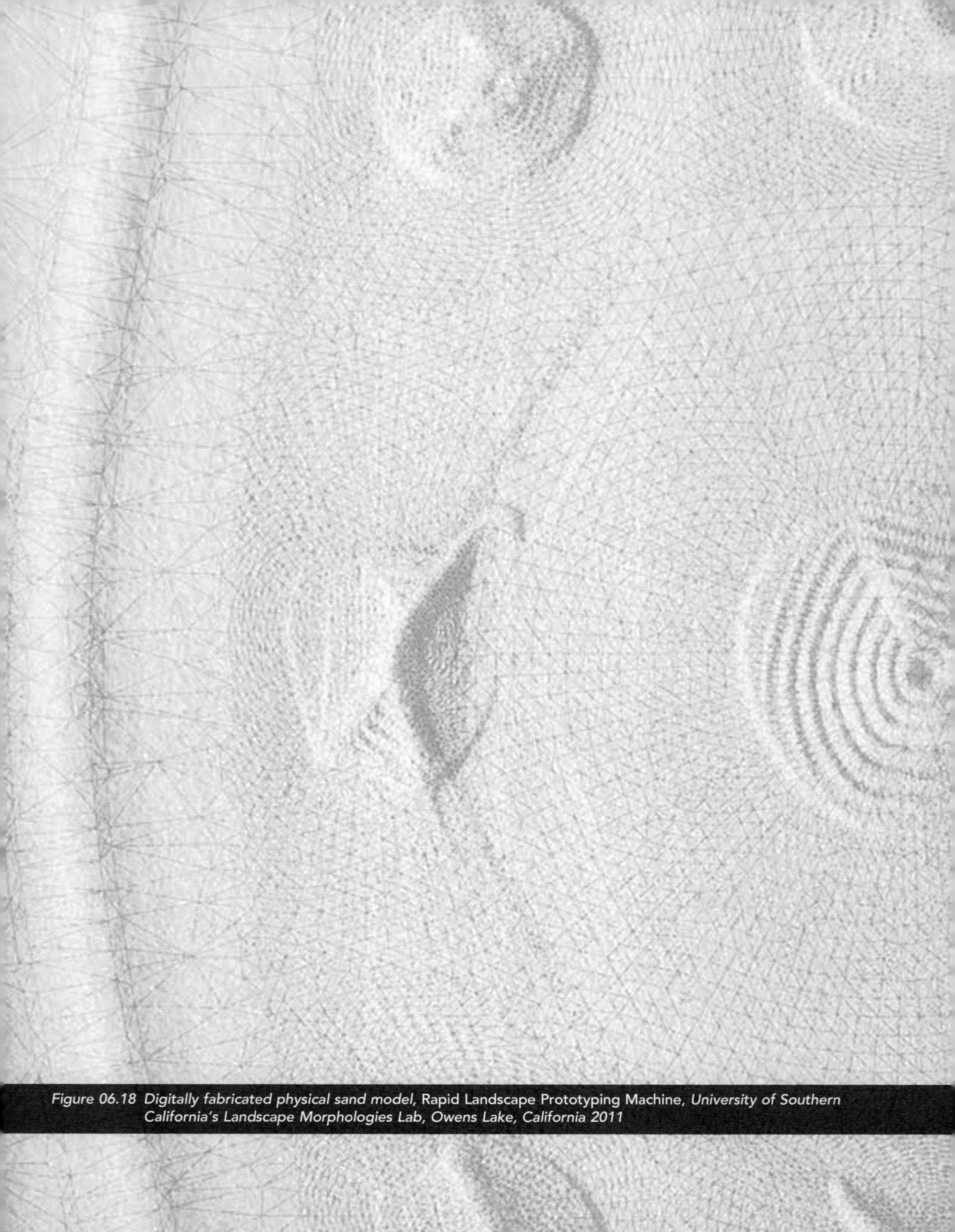

Figure 06.18 *Digitally fabricated physical sand model*, Rapid Landscape Prototyping Machine, *University of Southern California's Landscape Morphologies Lab, Owens Lake, California 2011*

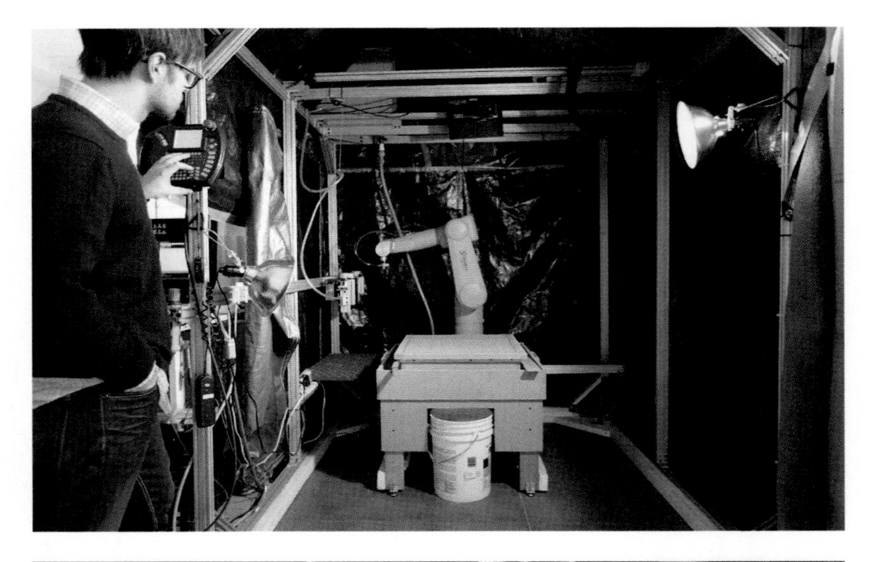

Figure 06.19
Custom fabrication lab, Rapid Landscape Prototyping Machine, *University of Southern California's Landscape Morphologies Lab, Owens Lake, California, 2011*

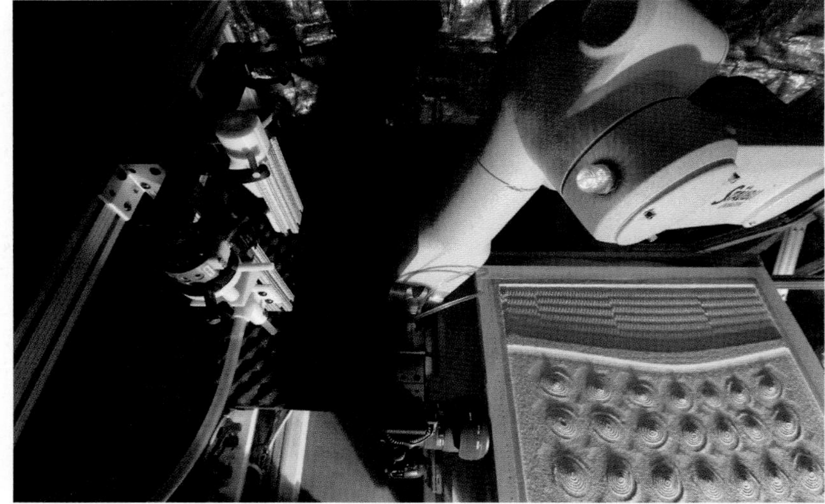

Figure 06.20
Robotic imaging of the Sand model, Rapid Landscape Prototyping Machine, *University of Southern California's Landscape Morphologies Lab, Owens Lake, California, 2011*

landscape.[39] The potential for translating complex computational tooling and patterning into earthen forms has been recognized. However, reductively carving a homogenous material to reveal surface textures fails to relate to the practices and performance of constructing landscape, resulting in a model that operates as an "image" rather than an expression or evaluation of a process oriented practice.[40] The RLPM bridges the ineffectiveness of current rapid prototyping efforts for landscapes to have relational material significance by employing the "effortless computational power and capacity of sand."[41]

As landscape architecture increasingly engages with large-scale disturbances, there is room to challenge civil engineering's "radical adherence to instrumental interventions" resulting in "unconsidered

Figure 06.21
Custom robotic end arm pieces, Rapid Landscape Prototyping Machine, University of Southern California's Landscape Morphologies Lab, Owens Lake, California, 2011

and unmeasured spatial consequences." Using a 3D scanner, Robinson is able to re-digitize the results of the sand modeling and "[enhance] the sand forms with morphologically-based adjustable surface treatments" with "real-time perceptual analysis tools."[42] Based off of methods commonly used within GIS to understand water levels, vegetation growth, aspect, and viewshed, the visualization of the analysis is projected back onto the Sand Model providing 3-Dimensional feedback. In addition, Robinson has created a game-like interface using the data from the 3D scan and subsequent analysis to navigate and render these conditions in real-time with given flexible parameters.

While the RLPM has been categorized within the chapter *Displace*, the project encompasses elements from the chapters *Elucidate* and *Compress*, expressed by the ability of the custom software suite to not only render but also project site-specific future landscape conditions. The simulations effectively portray seasonal water levels

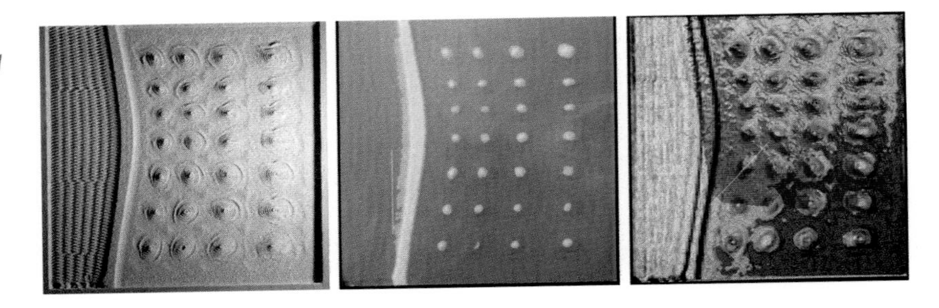

Figure 06.22
Analysis projections onto Sand models, Rapid Landscape Prototyping Machine, University of Southern California's Landscape Morphologies Lab, Owens Lake, California, 2011

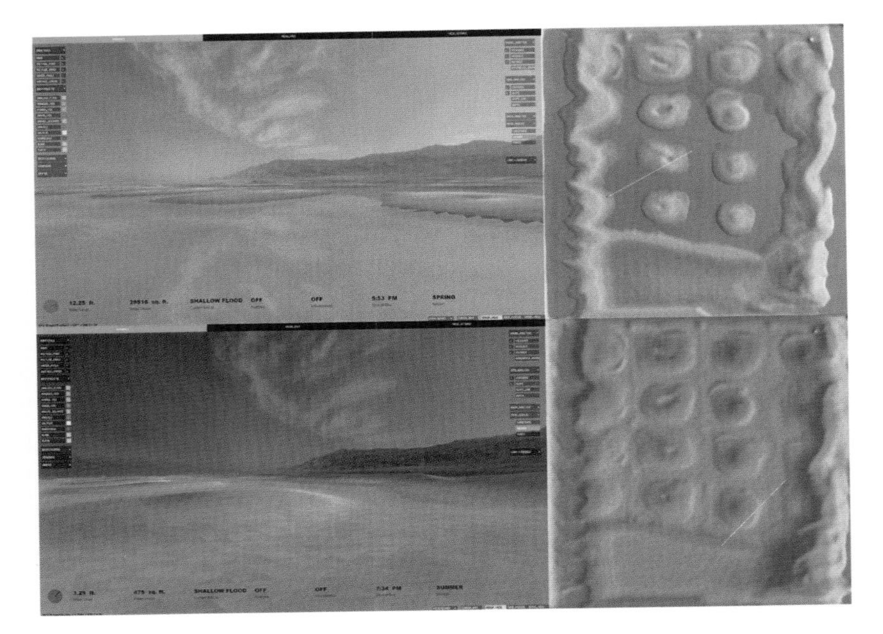

Figure 06.23
Real-time graphic simulation of Sand model landforms, Rapid Landscape Prototyping Machine, University of Southern California's Landscape Morphologies Lab, Owens Lake, California, 2011

and vegetative growth as well as daily and seasonal atmospheric conditions described by sunlight, shadow, and pigmentation of the sky and landscape.

This design tool becomes accessible beyond design professionals by linking together design methods with the representations required to facilitate design into a real-time feedback loop—allowing the user to visualize in real-time what would normally require a lot of post-processing. Robinson has streamlined these processes to craft experiential and analytical readings of landscape within a physically displaced location and predictive of possible future conditions—building a more immediate and illusory experience. Though sensory data is not being sourced in real-time from the actual landscape under investigation, this workflow is predictive of future working methodologies for process-driven dynamic landscapes, embracing a real-time connection with the landscapes they construct as a way of layering, nudging, and evolving ecological systems.

NOTES

1 James Carey, "Technology and ideology: the case of the telegraph," in *The New Media Theory Reader*, Eds. Robert Hassan and Julian Thomas (New York: Open University Press, 2006), 226. Previously published in *Communication as Culture*, London: Routledge, 1989

2 Ibid., 242.

3 Ibid.

4 Nicola Green, "On the Move: technology, mobility, and the mediation of social time and space," *The New Media Theory Reader*, Eds. Robert Hassan and Julian Thomas (New York: Open University Press, 2006), 250. Previously published in *The Information Society*, 18(4): 281–292.

5 Ibid., 251.

6 Denis Cosgrove, "The Measures of America," In Taking measures across the American landscape, edited by Corner, James, and Alex S. MacLean (New Haven: Yale University Press, 1996), 4.

7 Ibid.

8 Ibid.

9 James Corner, "The Agency of Mapping: Speculation, Critique and Invention," in *Mappings*, edited by Denis Cosgrove (London: Reaktion Books, 1999), 226.

10 Case Lance Brown and Rob Holmes, "Landscape Switching: A New Speed and Territory for Design Agency," *Kerb* 22, (2014): 44–49.

11 Jennifer Duggan, "Dead pigs floating in Chinese river," The Guardian (blog), April 17, 2014. http://theguardian.com/environment/chinas-choice/2014/apr/17/china-water.

12 Nicola Davison, "Rivers of blood: the dead pigs rotting in China's water supply," The Gaurdian (blog), March 29, 2013. http://theguardian.com/world/2013/mar/29/dead-pigs-china-water-supply.

13 Gary Strang, "Infrastructure as Landscape," *Places* 10, no. 3 (1996): 12.

14 Ellsworth and Kruse, *Making the Geologic Now*, 25 (see chap. 5, n. 10).

15 Timothy Morton, "Zero Landscapes in the Time of Hyperobjects," *Graz Architecture Magazine* 07 (2011): 78–87.

16 Henrik Ernstson, Sander E. van der Leeuw, Charles L. Redman, Douglas J. Meffert, George Davis, Christine Alfsen, Thomas Elmqvist. "Urban Transitions: On Urban Resilience and Human-Dominated Ecosystems." *AMBIO* 39, (2010): 537.

17 Ibid.

18 John May, "Logic of the Managerial Surface," *Praxis*, no. 13 (2011): 118.

19 Jesse LeCavalier, "All Those Numbers: Logistics, Territory and Walmart," *Places Journal*, May 2010. Accessed 11 Apr 2015. <https://placesjournal.org/article/all-those-numbers-logistics-territory-and-walmart/>

20 See Ian L. McHarg, *Design with nature*, Garden City, NY: Natural History Press, 1969.

21 Nicholas de Monchaux, *Spacesuit: fashioning Apollo*. (Cambridge, MA: MIT Press, 2011), 176.

22 Ibid.

23 Kristina Hill, "Shifting Sites," 145.

24 Burke, "Redefining Network Paradigms," 58 (see chap. 1, n. 37).

25 Khoury, Levit, Fong. http://khourylevitfong.com/

26 Donald Leopold, "Distortion of Olfactory Perception: Diagnosis and Treatment," *Chemical Senses* 27, no. 7 (2002), 612.

27 Nataly Gattegno and Jason Kelly Johnson, "Datagrove," Future Cities Lab Website, Accessed March 01, 2015, http://future-cities-lab.net/projects/#/datagrove/

28 Jason Kelly Johnson, "Thinking Things, Sensing Cities," In *Architecture In Formation: On the Nature of Information in Digital Architecture*, edited by Pablo Lorenzo-Eiroa and Aaron Sprecher, Routledge: New York, 2013.

29 Gattegno and Johnson, "Datagrove."

30 Franziska Eidner and Nadin Heinich. 2009. *Sensing space: Technologien für Architekturen der Zukunft = future architecture by technology*. Berlin: Jovis. (14)

31 Gordon Matta-Clark, Jeffrey Kastner, Sina Najafi, Frances Richard, and Jeffrey A Kroessler, *Odd Lots: Revisiting Gordon Matta-Clark's 'Fake Estates'*, New York: Cabinet Books and Queens Museum of Art and White Columns, 2005.

32 Nicholas de Monchaux, "Local Code: Real Estates," *Architectural Design* 80, no. 3 (May 2010), 90.

33 Ibid.

34 James Corner, "The Agency of Mapping: Speculation, Critique and Invention," in *Mappings*, Ed. Denis Cosgrove, (London: Reaktion Books, 1999), 213.

35 Nicholas de Monchaux, Local Code: 3,659 Proposals about Data, Design, and the Nature of Cities. New York: Princeton Architectural Press, 2015.

36 Jane Jacobs, The Death and Life of Great American Cities. New York: Random House, 1961.

37 Eric S. Raymond, The Cathedral & the Bazaar: Musings on Linux and Open Source by an Accidental Revolutionary. Beijing: O'Reilly, 1999.

38 Alexander Robinson, "Owens Lake Rapid Landscape Prototyping Machine: Reverse-Engineering Design Agency for Landscape Infrastructures," in *Paradigms in computing: making, machines, and models for design agency in architecture*, Eds. David Jason Gerber and Mariana Ibañez, (Los Angeles, CA: eVolo Press, 2014), 354.

39 May, "Logic of the Managerial Surface," 116–124.

40 Malcolm McCullough, Abstracting Craft: The Practiced Digital Hand. Cambridge, MA: MIT Press, 1996.

41 Robinson, "Owens Lake Rapid Landscape Prototyping Machine," 354.

42 Ibid., 351.

CONNECT

07

Connect facilitates direct interaction with the responsive system. The relationship it exhibits to the architecture, information, and feedback loop are apparent and easily interpretable. These projects provide a set of one-to-one relationships, where the inhabitant is cognizant of their connection and relationship to the overall system. Embedding responsive technologies in the landscape and the built environment open up new possibilities for connections. The field of responsive architecture has generally thought of these enhanced connections between user, technology, and site as a mediating device, particularly between public and private spaces. "Temporary or permanent, these techno-architectural interventions become mediating devices between the user, the artifact, and the site."[1] It can be argued that almost all forms of responsive architectures or landscapes craft some element of connection as a product of expanding the human computer interface. The ability to seamlessly connect attributes between elements focuses on modes of expression and drivers for previously unconnected systems. The underlying behavior is expressed through fundamental relationships, curated by the designer and appropriated to explicitly link phenomena or device to the user.

The projects outlined in *Connect* have particular agency for the development of connections that would otherwise be difficult to make by bridging physical, temporal, or perceptual barriers. Similar to the aforementioned methods within the chapters *Elucidate*, *Compress*, and *Displace*, methods of connection almost always require the translation of phenomena. Here the intention of translation exceeds interpretation and offers a language in which to communicate. The design of responsive systems within *Connect* requires some portion of participation for actuation, where the user is actively asked to engage with the system and prompt its responsivity through directed interaction. Many of Höweler and Yoon's responsive projects concerned with the public realm, perform through modes of connection "between users and space," in which they "investigate participatory environments, responsive architectures, and performative public spaces that combine architecture, technology, and landscape to propose new modes of publicness in the contemporary city."[2] Ideas surrounding the potential for responsive technologies to shape public interactions while mediating between ecological, public, and private spaces is perhaps the most central argument within the field of responsive architecture. This is illustrated by many of the projects in this book as well as the projects highlighted in *Interactive Architecture* and *Responsive Environments*.

The clearest methodology for designing connections within responsive landscapes is through direct one-to-one relationships, in which it is extremely clear how the participant's interactions prompt and engage responsive behavior. However, connections may also be formed through indirect relationships, in which phenomena are further abstracted. If connections are highly abstracted or require prolonged interaction, the method may be more related to projects within the next chapter, *Ambient*, where the connection is established over longer periods of time through contextual associations. *Connect* relies on an understanding that two or more elements, devices, or types of media are in relation to each other, and in which the participant is able to define a link. Once the connection and method of interaction are established, the participant continues to learn through interaction how to engage in the responsivity.

As a method, *Connect* implies that projects are latent, only activated through user interaction. The tactics employed to establish connections, whether they are direct or indirect, are formed in order to provide opportunities for interaction. First, methods are employed to capture the participant's attention, then offer known interactions. Meaning is then coupled with the interaction to ask the question: what is the consequence of this participation? *Connect* often asks people to act, to make slight decisions, whether it is a physical touch to operate devices, or how individuals move through space. Höweler and Yoon describe this as "[incorporating] its unscripted behavior into the project, relying on the interaction between the two to produce open source architecture."[3] Many of their projects exhibit direct connections that lend themselves to playful interaction. The value of supplying already known or easily learned interactions produces an interface that can be manipulated or engaged through playful interaction. Projects incorporating elements of play have been compared to musical instruments where the interaction is choreographed through actuation. These projects often have the intention of engaging the public in unique or novel social interactions that hinge on spatial or environmental qualities.

With the proliferation of electronic objects and digital artifacts, "We are experiencing a new kind of connection to our artifactual environment . . . partly visible, partly not."[4] Scenarios exhibiting *Connect* may be broad and apply numerous types of electronic objects and digital artifacts networked beyond in situ installations of responsive technologies. Humans interface with these technologies through telepresence, video chat, weather updates, and automated traffic lights; on top of that, countless actions are recorded, stocked, and stored away. The global positioning system in conjunction with urban and environmental monitoring, telecommunications, and distant data storage offer countless opportunities for connection and the synthesis of *big data*. As connections are heightened across networks of networks, multiple disciplines are quickly trying to understand the value and opportunities *big data* present. This interconnectivity and networking of information in urban environments often connotes conceptions of the city as "smart" or "efficient." Shannon Mattern states that "while the notion of the city as a data-generating, storing, processing, and formatting machine might not be new, the *reduction* of the city to those functions—which are increasingly *automated*—and the *reification* of that data, is distinct to our time."[5]

The digital networks and infrastructures of information sharing between objects and ideas are so connected, hierarchies of particular connections become impossible to decipher. The flat and dimensionless qualities of networked information have the potential to become rich with contextual information as interfaces reconnect to the landscape. *Connection* to the landscape has always been at our fingertips and within our field of vision. An extension of this relationship through responsive technologies has the potential to expand corporeal connections that are not entirely displaced, but networked. Anne Galloway describes "mixed-reality" as any combination of the physical and the virtual world to create "hybrid" environments, the most common being "augmented reality" and "augmented virtuality." The distinction is as follows: "Augmented reality seeks to enhance physical spaces and objects with virtual reality," while "augmented virtuality seeks to enhance virtual reality with real-world data and objects."[6]

In her essay "Big Nature," Jane Amidon asserts that "today, information and environmental technologies have the potential to virally increase awareness of ecological states, to link people, place, productivity, and performance."[7] Networked connections coupled with environmental technologies, particularly when extended to include the possibilities associated with augmented realities and augmented virtuality, have tremendous potential to address contemporary ecological issues. This potential is directly associated with the agency of social networks. In a paper titled "Urban Transitions: On Urban Resilience and Human-Dominated Ecosystems" Henrik Ernston et al. stress that despite the importance of social networks as shapers of ecological systems, they are fundamentally different from ecological networks. "Social networks are self-constructed by society in a process of 'alignment' or 'co-ordination,' best described as a continual recursive communication process."[8] Through continual correspondence and connection—intentions and information are exchanged—resulting in the movement of matter and energy. To Ernston et al., thermodynamic processes and material flows "could never have created durable human social institutions, let alone towns or societies."[9] Although social networks are fundamentally different from ecological networks, both urban and hinterland landscapes are "enacted and shaped through social networks,"[10] opening up opportunities for utilizing social networks to form connections within ecological systems.

Figure 07.02 Conceptual rendering, MIMMI, Invivia and Urbain DRC, 2013

MIMMI (Minneapolis Interactive Macro-Mood Installation)

Minneapolis, Illinois, 2013

INVIVIA and UrbainDRC

Collaborators: Allen Sayegh, Carl Koepcke, Jack Cochran, Bradley Cantrell, Artem Melikyan, Yuichiro Takeuchi

Scope: Inflatable and gathering area design and construction, software and hardware development

Technology: Custom web apps and language parsing of social media streams

---

MIMMI, the Minneapolis Interactive Macro-Mood Installation was an iconic inflatable cloud suspended above the plaza at the Minneapolis Convention Center Plaza. Cloud-like in concept, the inflatable hovered thirty feet above the ground gathering emotive information from Twitter feeds connected to Minneapolis residents and visitors to the plaza. It then analyzed this information in real-time to respond to the city's input with abstracted light displays and misting. *MIMMI* provided light shows during the evening and cooled the immediate micro-climates during the daytime. Whether the city was elated following a Minnesota Twins win or frustrated from the afternoon commute, *MIMMI* responded, changing its behavior in relation to daily human time cycles. Connection to *MIMMI* was established through visualization of both local user participation and externalized data streams. The physical public space was coupled with a far reaching network that collectively aggregated the urban experience, a connection that was typically experienced as a single viewer.

MIMMI was conceived as an emotional gateway to Minneapolis, bringing residents and visitors together to experience the collective mood of the city. Seeking to engage both the virtual and physical layers of the community, *MIMMI* used technology to see the city in a new way and also to reinforce the serendipitous gathering that characterizes urban life. The design team proposed *MIMMI* as a productive response to how cities and societies have evolved in relation to ubiquitous digital media, taking advantage of the new opportunities and insights such technology provides while working to balance those privileges with new responsibilities as the cities change. The abstracted and ubiquitous data from the larger population is connected to the behaviors of MIMMI and visualized for a local audience. In this sense, *MIMMI* merged the discussions of digital technology and a densifying urban environment, ultimately creating a place to gather and see the city in a new way, while at the same time exploring shifting cultures and responsibilities.

To generate the city's mood, *MIMMI* sourced information from local Twitter feeds and used textual analysis to detect the emotion of those

Figure 07.03 Night view of inflatable misting, MIMMI, *Invivia and Urbain DRC, 2013*

Figure 07.04 Melancholy evening mood, MIMMI, *Invivia and Urbain DRC, 2013*

tweets. By aggregating the positivity and negativity of tweets in real-time, *MIMMI* connected the abstracted emotion of the city to a series of Wi-Fi-enabled LED bulbs and integrated misting system. Each of these visualization methods interfaced the external city and local site with dual modes of user interaction. The low-energy lights, hung inside the balloon material and stretching throughout the entire shape, displayed the mood—beginning at sunset. The color of the lights

Figure 07.05
Daytime view of plaza area, MIMMI, Invivia and Urbain DRC, 2013

shifted from cold colors (negative) to warm and hot colors (positive) depending on the mood, with a rate of change dependent on the rate of tweets. Water activity occurred during the day through tubing and nozzles embedded in the fabric of the balloon, with higher levels of misting occurring when the city was happier.

Visitors to the plaza formed an integral part of *MIMMI*'s behavior, as they interacted with *MIMMI* and helped improve its mood. If the city was particularly *sad* or emotional, citizens could come together to lift *MIMMI*'s spirits as *MIMMI* could detect movement at the plaza and include this information in its analytics. The more people present under the cloud and moving around, the more active *MIMMI* became, responding either with increased lighting or misting, depending on the time of day. This localized interaction was connected virtually to the aggregated mood of the city and had the ability to transform *MIMMI*'s behavior.

In addition to the physical installation, a website was developed. The website, part of the Minneapolis Convention Center's site, catalogued the mood of the city generated by *MIMMI* over the summer of 2013, allowing visitors to see daily and weekly trends in the city's emotions.

The website also featured links to events at the plaza and the Convention Center. Visitors could also tweet directly to *MIMMI* using the hashtag *#mimmi*, having an influence over the aggregate mood of the urban population.

The *MIMMI* installation explored very specific connections between large-scale social behavior and localized interactions. The translation of a populace's mood provided interesting insights between how people communicate and the language used to describe particular events. Similar projects such as the *Hedonometer* attempts to evaluate collective mood through language parsing and visualize this relationship in a web-based interface.[11] Jonathan Harris and Sepandar Kamvar also explore mood through their web-based project, "We Feel Fine," which attempts to visualize the mood of the internet by parsing the language used in blogs and other posts. "The interface to this data is a self-organizing particle system, where each particle represents a single feeling posted by a single individual. The particles' properties—color, size, shape, opacity—indicate the nature of the feeling inside, and any particle can be clicked to reveal the full sentence or photograph it contains."[12] *MIMMI* used these projects as inspiration and began to attach similar data to the landscape and spatial interactions. Operating as an icon within the city, the installation tested methods of interaction based on connections between multiple scales and socio-cultural spaces.[13]

Interview with Allen Sayegh

Invivia

Allen Sayegh is an architect, designer, an educator and the principal of INVIVIA – an award winning global design firm. He is an Associate Professor at Harvard Graduate School of Design and the director of REAL the Responsive Environment and Artifacts Lab at Harvard. Sayegh's academic research and professional practice span a period of two decades working on projects of varied scales. He has taught at different institutions and has worked on many different projects throughout the US, Europe, the Middle East, and Asia.

01

The work at Invivia is broad, exploring media and visualization, developing installations that sense and respond, and designing space and experience. With this breadth of work, what are the major influences for Invivia? What are the larger research objectives for the firm?

For me it is important to experiment with very different types of situations and media; through experimentation, it helps each project come together. I can see that people say we are doing many different types of things but I see them all feeding into a similar idea. It is basically about finding opportunities with technology to create new types of experience that happen to be in the built environment. If you are designing interaction, you need to understand behavior, psychology, and many other aspects. We are in a time where professions are transitioning and we are going to have many hybrid disciplines such as engineer animators or design scientists. This is part of a transition, a hybrid state that myself and many others are experiencing. You see, in Silicon Valley the need for positions such as a design engineer, people want that kind of talent. There is a need for fluidity between knowledge areas and this forms new disciplines over time.

02

You have used the term "highly evolved useless things" to encapsulate many of the prototypes and installations Invivia has developed. What would you consider a highly evolved useless thing? How does this concept help to push work in responsive technologies forward?

I think that is half of the sentence, I would say "highly evolved useless things that have an evocative power." It is a form of experimentation that is evocative and can lead to multi-layered complexity. The "useless" aspect is tongue in cheek, it has to have a power, similar to a function. It is a counterpoint to the continual need for efficiency that dominates computation and instead evokes something that is more whimsical and interesting. Humans have a range of needs beyond efficiency, we are not robots. The highly evolved useless thing speaks to cultural and social intangibles.

03

The work of responsive projects such as *Tulipomania*, the *Einstein Gravity Wall*, and *MIMMI* have focused primarily on exploiting a highly crafted effect. This high level of craft not only focuses on a physical object, but also on the expression of an experience. Do you see similarities in crafting the physical and the virtual? Are there major differences between the two?

I don't see that there are major differences. I approach it in the same way. It is more about finding an effect that blends the two while being critical of why it is physical or digital. Once you understand the potentials of each you can highlight the effect; there is nothing worse than doing something digitally that could be better performed physically and vice versa. You just need to know the limits. The one issue with the digital is that it changes so rapidly that there are new potentials that always come up, these boundaries keep changing and you have to understand the time in which it is being built. Fundamentally it is the same, the change with the digital is that the pace is much faster. If you are doing visualization and rendering you have to constantly learn new things, this isn't the same with sculpting marble. With digital the technique is changing and evolving faster than it is possible to become an expert. So in that way there are no experts, everyone is experimenting. You have to become an expert in adapting and evolving.

04

At Invivia, many projects focus on the intangible or subjective qualities of human experience. This doesn't always align with the objective qualities of digital or responsive technologies. Do you find that there are contradictions and in what ways is this approach beneficial?

I don't find that there are contradictions, I think it is actually complementary. There might be a tension, but many times you are attempting to transcend that and it creates the project. The project relies on interesting uses of the technology that are attached to human experiences. A current project we are developing in Copenhagen is an installation about empathy, it is poetic.

05

What technologies are pushing the work at Invivia? Are there particular developments in responsive technologies that interest you moving forward?

I am not interested in anything specific. Whatever is the latest thing, we experiment with it. But there is no need to concentrate on a specific technology. There is so much change that it isn't possible to be immersed into a single technology.

06

How do your connections with academia feed into the work at Invivia? In the reverse, how do you see Invivia's work feeding back into your teaching or academic research?

I see it as one thing, I have been teaching for so long that they are difficult to separate. There is a slowness to academic research that affords time for speculation, which can be good. In some ways the professional work is more cutting edge though and the academy is attempting to catch up. There is a pace in professional work that requires decision making and always a need to make things work this complements the more speculative work in the academy.

Aviary
Dubai, United Arab Emirates, 2013

Höweler + Yoon and Parallel Development

Collaborators: Audio composition by Erik Carlson, Computer programming by Jason Cipriani and Nicholas Joliat

Scope: Two small-scale permanent interactive and responsive light and sound installations

Technology: Capacitive Sensing

Aviary is a small-scale public installation of light poles designed by Höweler + Yoon and Parallel Development, which reacts to people's touch with light and sound to perform and be played like birds in their natural habitat. Situated in the Dubai Mall, *Aviary* is composed of two like installations, each forming a spiral. The installations require direct connection with people in order to form a response. The response of light and sound is intended to be played like a musical instrument. The interaction and connection to the response is immediate—establishing a feedback loop between the installation and the "player" leading to informal but informed real-time performances of light and sound.

Aviary is an evolution of an earlier project by Höweler + Yoon, titled *HI FI*, a public-scaled musical instrument installed on a sidewalk in Washington DC composed of twenty touch-activated sound poles "using the human touch to create a responsive environment of soft, sonic envelopes, the public activates the field and engages in play and performance."[14] *Aviary* utilizes sophisticated technological applications to create a highly sensitive and networked interactive experience. This method of interfacing with responsive audio-visual light poles can be traced to an installation designed by Christian Moeller in 1977 titled *Audio Grove*—a field of similarly interactive light poles positioned to encourage spontaneous and interactive spatial compositions of movement, light, and sound—commissioned by Spiral Gallery/Wacoal Art Center in Tokyo and exhibited at multiple locations.[15] The design of *Aviary* has evolved earlier precedents by expanding the interaction to embody a behavioral narrative of avian species.

The sound composition created by Erik Carlson shifts between synthesized digital tones at the base of the pole and life-like recordings of songbirds towards the top of the poles. This transition invokes the metaphorical use of songbird behaviors to facilitate human interaction within a synthetic responsive installation. The collection of bird species sampled for the compositions are either native to or pass through the Arabian Peninsula region, which serves as the migration route between

Figure 07.06 Capacitive sensitive sensing with silver nanowire conductive layer, Aviary, Höweler + Yoon and Parallel Development, Dubai Mall, Dubai, United Arab Emirates, 2013. Courtesy of Höweler + Yoon Architecture

Figure 07.07 Aviary, *Höweler + Yoon and Parallel Development, Dubai Mall, Dubai, United Arab Emirates, 2013. Courtesy of Höweler + Yoon Architecture*

Africa and Asia. Each pole represents and holds access to a particular bird species within a network of poles that synthesize birds flying in song.

The choreography of light and sound is facilitated by a direct connection between the user and physical interaction with the poles through various methods of touch. The technological components of each light pole include 96 "touch points" activated through capacitive sensitive sensing of the body's electrical field, encased in a cylinder of clear film with a silver nanowire conductive layer, then processed by a small computer located at the base of each pole to actuate full color LED lights and speakers. A short touch will produce a dim localized light and a sustained hold will fill a localized area of the pole with light. Sliding or swiping along the pole in the vertical direction upwards sends light up the pole and across to other poles increasing speed results in corresponding sound effects. A quick slide down will deliver a "migration" of light and sound across multiple poles mimicking the collective conversing of birds within an aviary.

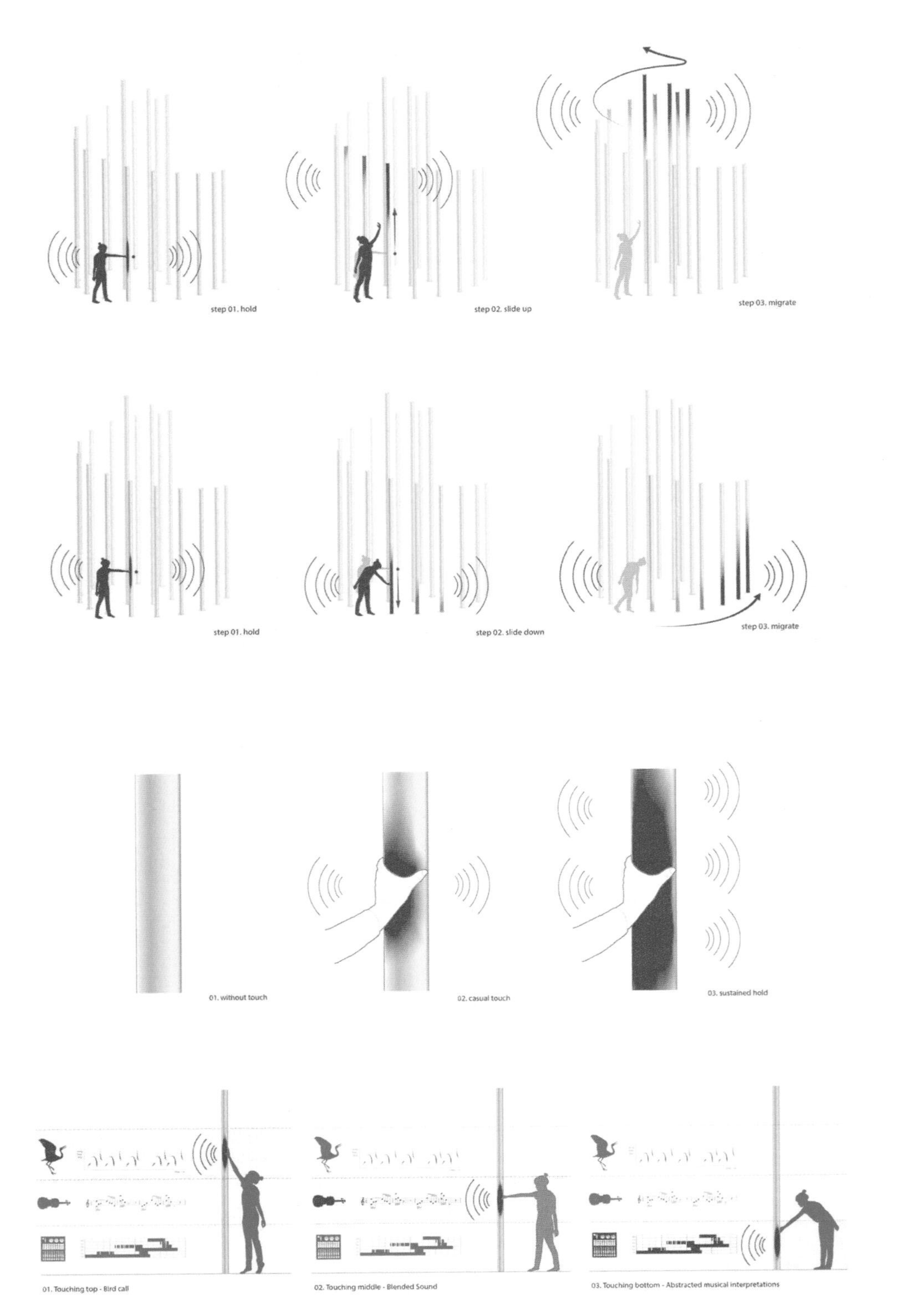

Figure 07.08 *Interaction and sound diagram, Aviary, Höweler + Yoon and Parallel Development, Dubai Mall, Dubai, United Arab Emirates, 2013. Courtesy of Höweler + Yoon Architecture*

The directness of these relationships facilitates an interaction between user and interface; although sophisticated, its operation easily becomes transparent, encouraging playful and experimental interactions. Creating interfaces with logical methods for stimulating actuation forges learned interactions leading to known or predictable behaviors defined by the established connections.

Figure 07.09
Aviary, Höweler + Yoon and Parallel Development, Dubai Mall, Dubai, United Arab Emirates, 2013. Courtesy of Höweler + Yoon Architecture

Figure 07.10 Sky Ear, *Haque Design + Research, Fribourg, Switzerland and Greenwich, London, 2004*

Sky Ear

Fribourg, Switzerland and Greenwich, London, UK, 2004

Usman Haque, Haque Design + Research

Collaborators: Seth Garlock, Senseinate Inc. Electronic Engineer, and Rolf Pixley, Anomalous Research, for software development. Carbon fibre tubing for framework by RBJ Plastics, UK

Scope: One night responsive installation

Technology: Cell phones, Infrared Sensors, LED lights

Sky Ear, designed by Usman Haque, is a one night installation at the Belluard Bollwerk International Festival, Fribourg, Switzerland, and the National Maritime Museum, Greenwich, London, in 2004 to explore and connect people to the "Hertzian landscape"—termed by Anthony Dunne to describe the increasing electromagnetic environment created by electronic objects.[16] Funded by The Daniel Langlois Foundation for Art, Science, and Technology, *Sky Ear* is a *cloud* of glowing balloons and cell phones, suspended in the sky to sense and visually display the present and relative electromagnetic phenomena of ubiquitous cellphone networks. Electromagnetic fields are invisible to human senses, *Sky Ear* renders the phenomena while prompting interaction and connection. Viewers are asked to call into the cloud and listen to the sounds of the sky and the electromagnetic field above. As participants call and text message the cell phones suspended within the cloud, their interaction actively alters the electromagnetic phenomena, "the cloud is actually responding to the electromagnetic fields created by the phones in the cloud."[17]

Figure 07.11 Sky Ear, *Haque Design + Research, Fribourg, Switzerland and Greenwich, London, 2004*

Heavily influenced by the work of cybernetician Gordon Pask, particularly his theories on "underspecified and observer-constructed goals," Haque extends and tests these concepts through the development of interactive and responsive systems in architecture.[18] The concept for *Sky Ear* emerges from Haque's personal research of "hard" and "soft" space in relation to the movement of cell phones from hardlines, exploring "how mobile phones condition our use of space now that they have become ubiquitous."[19] As well as the invisible and seemingly immaterial qualities of electromagnetic field emanating from the discrete object, the cell phone. Haque is influenced by the work of Anthony Dunne and his exploration what he calls "radiogenic objects," that is an object designed to "function as unwitting interfaces between the abstract space of electromagnetism and the material cultures of everyday life" in order to link "the perceptible material world to the extrasensory world of radiation and energy."[20]

Sky Ear is structured with a metal armature connecting one-thousand extra-large helium balloons into a large flexible mattress. Each individual balloon is equipped with six ultra-bright LEDs capable of producing an infinite range of colors. The color display responds to electromagnetic inputs collected from the surrounding air and altered by the activity of the cloud "like a glowing jellyfish sampling the electromagnetic spectrum rather like a vertical radar sweep."[21]

Figure 07.12
Sky Ear, *Haque Design + Research, Fribourg, Switzerland and Greenwich, London, 2004*

Figure 07.13
Sky Ear, *Haque Design + Research, Fribourg, Switzerland and Greenwich, London, 2004*

The electromagnetic signals contributing to the rendered landscape above are produced from multiple sources. Although the visualization prioritizes fields originating from mobile phones, the sensors pick up distant storms, radios, broadcasting, and other forms of electromagnetic radiation. Furthering the ability to capture this ephemeral landscape phenomena, the balloons are able to emit and receive signals across the armature through infrared sensing—elucidating connections across the sky as a larger field condition.

In considering this form of interaction, *Sky Ear* utilizes a ubiquitous form of connection between people—the cell phone. This method of connection by phone call or text message from person to person is conceptually clear but the physical manifestation of the connection is imperceivable. *Sky Ear* visualizes the physical connection by representing the resulting phenomena while still providing a known connection to the sounds of the sky, each feeding off the other. The impressive scale of the installation, landscape scale, provides a unique spatial experience of the "Hertzian landscape," one that speaks to its reach and extent. The interaction with *Sky Ear* provokes phone calls, asking the participants to augment the electromagnetic field, "humanizing Hertzian space" in which the installation becomes less about the visual spectacle, and more about an active performance, "encouraging the poetic and multi-layered coupling of electromagnetic and material elements to produce new levels of cultural complexity," and forging connections between the perceptual and imperceptual landscape.[22]

Figure 07.14 Amphibious Architecture, *The Living and The xClinic Environmental Health Clinic, New York City, 2009*

Amphibious Architecture

New York City, 2009

The Living Architecture Lab at Columbia and The xClinic Environmental Health Clinic at New York University

Collaborators: Mark Bain, David Benjamin, Amelia Black, Natalie Jeremijenko, Abha Kataria, Jonathan Laventhol, Deborah Richards, Zenon Tech-Czarny, Kevin Wei, Chris Woebken, Soo-In Yang

Volunteers: Steven Garcia, Yong Julee, Donghyuk Chang, Jamie Kim, Laura Vincent, Seong Kyn Han, James Yoon, Anastasiya Konopitskaya, Anna Sawicka (Anka), Kianga Ford, Thomas Stroman, Jörg Thöne, Junho Cho, Moon Young Kim, Kyung Jae Kim, Brian Sweeney, Zachary Colbert, Eva Lansberry, Stephen Man, Adam Brown, Daehee Lee, Jung Gu Kim, Jorden Lu, Younjung Kim, Sidharth Kadian, Steve Sanderson, Cathy Jones, E Harper Jeremijenko Conley, Yo Jeremijenko Conley, Dalton Conley

Sponsors: Arup Bronx River Alliance, Bronx River Art Center, J.A. McDermott Lighting Corp., New York City Economic Development Corporation, New York University Department of Art, New York University Computer Science Department, Pachube, SHoP Architects, Silver Screen Marine, Studio-X at Columbia GSAPP

Scope: Two Temporary Installations on the Brooklyn River and East River in New York City (40′ x 40′ x 6′) connected with an SMS Interface

Technology: Arduino IDE, Lo-fi Fish Sensors, LEDs, SMS Interface

---

Amphibious Architecture is a unique collaboration between Soo-In Yang and David Benjamin, director of The Living; Natalie Jeremijenko of the Environmental Health Clinic; and Chris Woebken, co-founder of The Extrapolation Factory. With support from the Architectural League of New York, *Amphibious Architecture* was exhibited as part of *Toward the Sentient City*, curated by Marc Fornes to explore the role of networked technologies to shape near-future urban environments. The Living Architecture Lab is a speculative lab at Columbia that explores new relations and applications between architecture and responsive technology. The Environmental Health Clinic addresses environmental issues through the "inversion of health as an external phenomenon"[23] by using a combination of science, technology, and environmental art. Lastly, The Extrapolation Factory is recently known for his work with responsive technologies and animal sensing capabilities. Together they designed *Amphibious Architecture*, a floating light installation situated on the Bronx and East River in New York City performing as a physical and digital "two-way interface" to bridge ecosystem interactions between land and water "as a new kind of public spectacle."[24] This installation asks how can ubiquitous computing and in situ monitoring be coupled to render the shared environment of urban river ecologies?

Figure 07.15
Amphibious Architecture,
The Living and The xClinic Environmental Health Clinic, New York City, 2009

Rather than elucidate information surrounding water quality and eco system health, such as displaying dissolved oxygen levels (already highly monitored by NOAA) as a data-visualization, the designers chose to use the localized presence of fish as indicators of ecological activity and subsequent environmental health. In an interview with Geoff Manaugh, Chris Woebken describes re-purposing cheap lo-fi battery powered fish sensors designed for remote fishing to detect different fish species in the river. The data from this localized network of lo-fi sensors was utilized to *connect* visitors and fish in real-time to "create a voice for the animal" and inform an SMS interface for visitors to text and receive responses from the fish about "what's going on."[25]

Each installation is composed of floating components, built to encase sensing and processing technologies to actuate LED light displays both above and below the water's surface. The light display above the

Figure 07.16 SMS interface, Amphibious Architecture, *The Living and The xClinic Environmental Health Clinic, New York City, 2009*

surface tracks water quality over time, glowing with warm purples and pinks if the water quality is worse than the previous week or with cool blues if the water quality has improved. Below the surface, the lights trace the paths of fish swimming below the installation, spatializing their normally invisible presence. As citizens text the interface, the lights blink twice, actively marking participation and physically rendering communications across networked devices. The system updates once a second, portraying the system as truly dynamic and capable of engaging in conversation related to human temporal scales of information exchange.

Fish (unlike dissolved oxygen or nutrient levels) are both spatially and emotionally relatable; not only can citizens visualize the fish, but they also trigger empathy and compassion. Attaching ecological issues to non-human organisms to affect change has been a strategic tool for conservation and preservation efforts. These "charismatic megafauna" or celebrity species (such as the giant panda or bald eagle) are used as leverage to shape public opinion about related ecological issues, sometimes finding their way into ecological design proposals. In a way, *Amphibious Architecture* has evolved this strategy to expand our view of urban environments and ecological systems through customized connections with other species "to create a dialogue where people could talk to the water—to the organisms in the water."[26]

Figure 07.17 *Prototype*, Amphibious Architecture, *The Living and The xClinic Environmental Health Clinic, New York City, 2009*

In addition to drawing attention to water quality issues, the goal of the project was to "engage citizens to think of the city as an open-ended collective experiment"[27] by bridging perceptual and communication barriers between underwater environments and the habitable urban landscape. Extending communications to fish and aquatic organisms is just the beginning of potential forms of communication through responsive technologies. These technologies evolve the translation of sensed information to fit into known methods of social interaction—in this case, texting. This form of connection draws upon how we are already using networked devices—ubiquitous computing—and communication networks, and opens up potentials for harnessing existing infrastructure to find novel more potent forms of connection to environmental conditions.

Figure 07.18 *H.O.R.T.U.S, Marco Poletto and Claudia Pasquero, ecoLogic Studio, AA Front Members Room, London, 2012*

H.O.R.T.U.S
London, UK, 2012

ecoLogic Studio

Collaborators: Claudia Pasquero and Marco Poletto

Scope: Temporary installation of responsive algae garden for the AA Front Members Room in London

Technology: Arduino, Piezo Buzzers, Proximity Sensors

H.O.R.T.U.S—an indoor intelligent and synthetic garden by ecoLogicStudio—is an acronym for "Hydro Organisms Responsive To Urban Stimuli," and is a nod to the etymology of the word "garden" from the German "Garten," which originally meant "enclosed or bounded space." This, when translated to Latin, reads "HORTUS conclusus."[28] ecoLogicStudio is an architectural and urban design studio co-founded by Claudia Pasquero and Marco Poletto, "defined by the combination and integration of systemic thinking, bio and socio-logic research, parametric design and prototyping."[29] They have taken a specific interest in what they call the synthesis of "agri-urban" systems—bridging the speed of digital information flows with the speed of material and biological processes through the use of responsive technologies to experience and adjust these systems in real-time.[30] *H.O.R.T.U.S* physically ties a human process and a landscape process together.

Figure 07.19
H.O.R.T.U.S, *Marco Poletto and Claudia Pasquero, ecoLogic Studio, AA Front Members Room, London, 2012*

The physical installation of *H.O.R.T.U.S* is composed of plastic pouches hanging from the ceiling around eye level filled with multiple species of micro and macro algae, all differing in color and texture. Each algae bag or "proto-garden" is connected to a tube for the viewer to breathe out into and inject carbon dioxide into the container—increasing the nutrient input and therefore increasing the oxygen output by the algae into the room. Groups of algae bags are connected to larger containers or nodes holding bioluminescent bacteria. As oxygen builds up within the algae bags, a pump is actuated that distributes the oxygen to the bioluminescent bacteria containers. Ambient light sensors are used to detect the light of the bioluminescent bacteria and asses the performance of the overall garden.

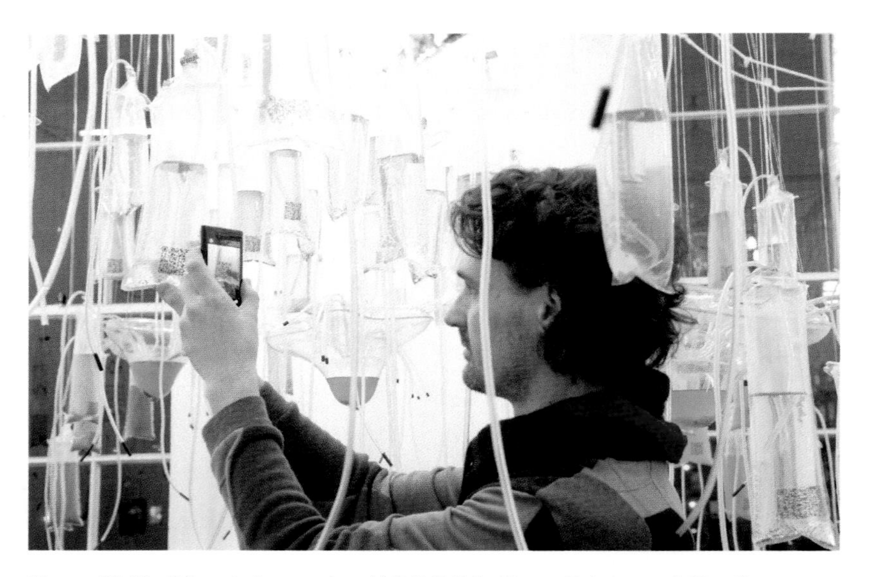

Figure 07.20 QR code interaction, H.O.R.T.U.S, *Marco Poletto and Claudia Pasquero, ecoLogic Studio, AA Front Members Room, London, 2012*

Termed "cyber-garden," the proto-gardens are networked both physically and digitally. The physical installation is coupled with a virtual garden, tracking and compiling scans and tweets surrounding visitor interactions and perceptions. Each pouch is equipped with a QR code. After feeding the algae with carbon dioxide, the visitor can scan the QR code with their smartphone to view real-time data on a web page about the specific algae bag they are interacting with. The concept of "cyber-gardening" draws from landscape practices, specifically gardening, as a method for maintenance practices in which the garden is observed and tended to over time. They were particularly inspired by Gilles Clément—a French landscape architect, ecologist, botanist, and entomologist—whose work explores "humanist ecology"

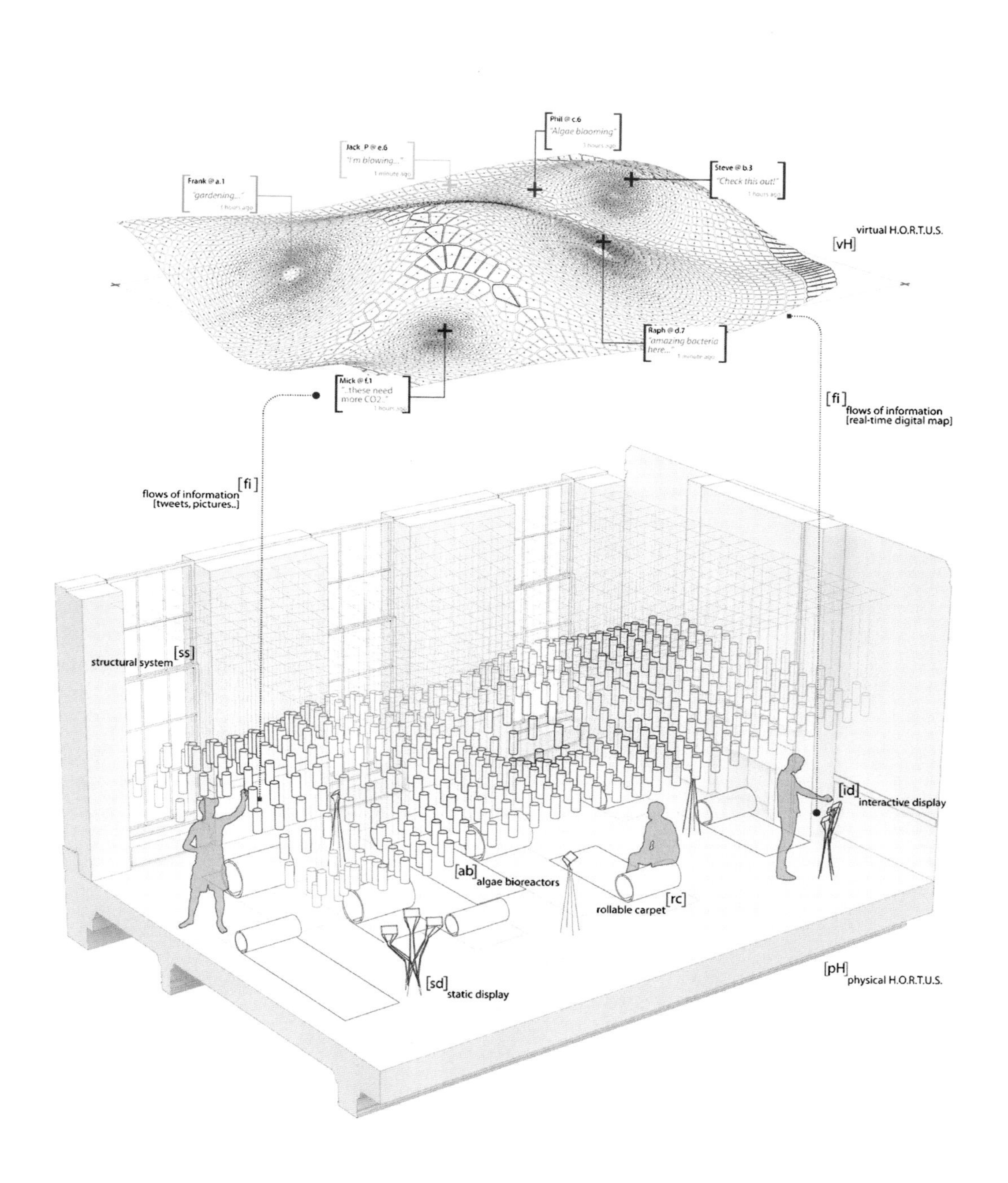

Figure 07.21 Installation structure and operation, H.O.R.T.U.S, Marco Poletto and Claudia Pasquero, ecoLogic Studio, AA Front Members Room, London, 2012

and "biological gardening" as themes for investigating concepts of wildness, growth, and decay within artificial practices of gardening.[31] *H.O.R.T.U.S* functions as a miniature laboratory designed for new technological gardening methods as well as for close inspection. The field of samples perform as micro experiments in which the participants become the monitors as well as gardeners. The installation was maintained by both physical flows of energy and matter and digital flows of information during its four-week growing period, "inducing multiple mechanisms of self-regulation and evolving novel forms of self-organisation."[32]

H.O.R.T.U.S speaks to the potential for highly monitored social engagement with ecological systems to increase participation within productive and working landscapes. This participation increases support for urban landscape conditions and introduces possibilities for integrating productive landscapes into urban environments. ecoLogic Studio's project *Regional Algae Farm*, devised for the Swedish Region of Osterlen, imagines the utility of unique algae based systems across multiple sites within the region. Referred to as an "interface" of networked sites rather than a masterplan, their proposal highlights the intent of participatory design to employ both top-down and bottom-up strategies that involves local stakeholders, while at the same time considering the performance of regional systems: "models need testing and feedback from people, so that prototypes can be optimally identified as multi-use educational resources related to their living context."[33] The design of *H.O.R.T.U.S* functions as a diagram and prototype for large-scale interventions. It offers methods based on novel intimacy with scalar landscape systems through social media and networked devices with the objective of promoting citizen participation to actively shape synthetic ecologies.

Keiichi Matsuda. Critical Design

Scope: 3:00 Minute Video

Technology: Motion Graphics and Video Editing

The idea of virtual reality can be traced back to Myron Krueger's use of the term "artificial reality" in the 1970s, popularized by movies such as *Lawnmower Man* in the 1980s.[34] This other place, a virtualization that convinces our senses of a *legitimate* experience can be found in a range of hybrids outside of the virtual realities that require headsets or immersive spaces. In practice, this virtual space exists in our media, video games, advertisements, and film. These realities are packaged into devices that sit in our living rooms, hide in our handbags, and reside on our office desks. Rather than diving deeper into these devices, augmented reality turns this relationship inside out and embeds media into the physical environment. The concept embraces an overlay of media, spatially positioned to correspond with the physical world. This develops relationships between the material world and the virtual, syncing simulation-time and physical-time to create a rich hybrid experience. *Augmented City 3D*, a film by Keiichi Matsuda, dives into this concept and imagines a world where the non-material is as valid as the material, often obfuscating material with layers of media.

Augmented City 3D is the second in a series of films by Matsuda that examine the world from the lens of an augmented reality interface, speculating how our relationship with the material world would be altered. Preceding *Augmented City 3D*, Matsuda's film *Domestic Robocop* illustrates the future home as a "dislocated domestic construct merged seamlessly with the augmented city."[35] Both films present a world that is open, while at the same time contradictorily optimistic and dystopian. The films explore a fundamental relationship between public and private space, particularly the insertion of corporate space in historically personal moments. Matsuda re-appropriates the discussion of public and private using terms such as "broadcast" and "aggregation," which more readily align with a discussion of augmented spaces.[36] *Augmented City 3D* envisions a future of customization where users overlay veneers to "style" space with spatialized content.

The film frames the built environment as an infrastructure or scaffold decorated with overlays of media. The coffee shop and the street become the canvas for simulated paint with unlimited possibilities where architects are freed from the constraints of physics and are able

Figure 07.22 Conceptual diagram, Augmented City 3D, *Keiichi Matsuda, London, UK, 2010*

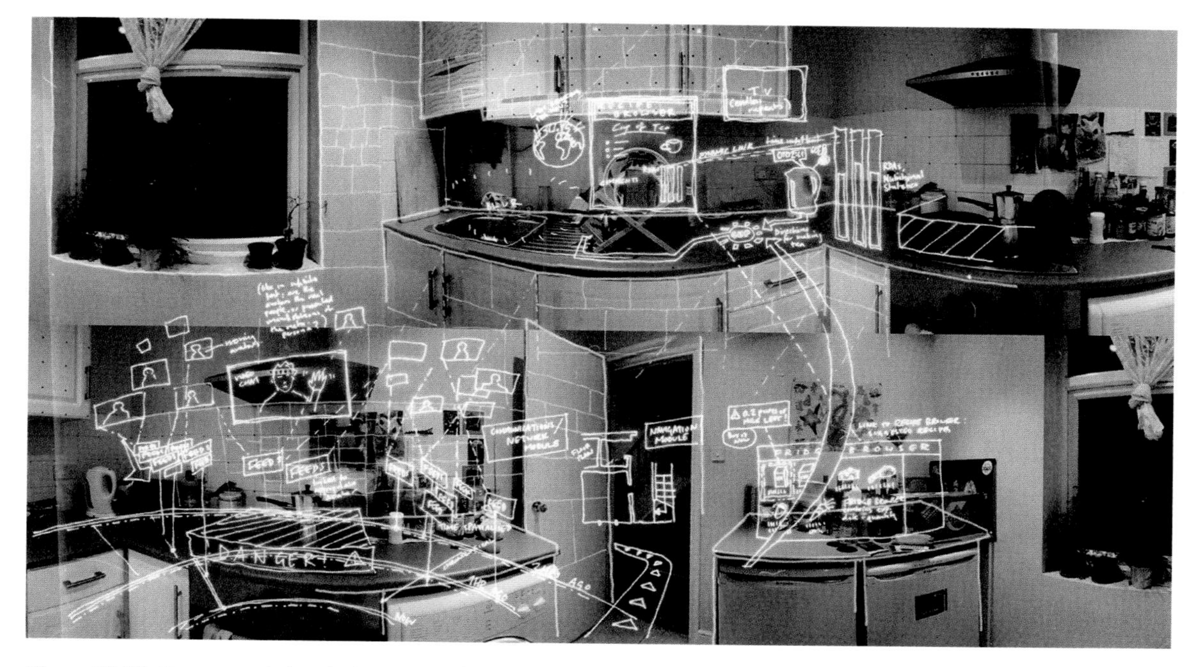

Figure 07.23 Conceptual sketch, Augmented City 3D, *Keiichi Matsuda, London, UK, 2010*

Figure 07.24
Animation process screenshot, Augmented City 3D, *Keiichi Matsuda, London, UK, 2010*

to fully sculpt the landscape from the human eye out. Matsuda depicts his speculations through hybrids, mixing the existing physical world with media overlays. The insertion of corporate advertising and personalized commercial offerings inspect how this future might be commandeered. Interestingly, the film does not examine the implications of this form of augmented reality on the material of infrastructural qualities of the landscape. Instead, it insinuates that the physical realm slowly becomes redundant, unnecessary in a space where the ideal can be projected: a reality in which the underlying infrastructure is merely a canvas, where individual architectural interests can be dialed in and customized by the user. Perhaps an architecture defined as a post-modern bricolage, subsumed by personal kitch and corporate advertising.

Figure 07.25
Urban interface, Augmented City 3D, *Keiichi Matsuda, London, UK, 2010*

The film is an important investigation into landscape as the underlay for a future of augmented realities. While we rush to embed media into all aspects of the city, home, and workplace, *Augmented City 3D* provides a baseline to view how dreams of technological democratization, personal customization, and real-time data can lead to further displacement through a feigned connectivity.

Figure 07.26
Driving interface, Augmented City 3D, *Keiichi Matsuda, London, UK, 2010*

Interview with Keiichi Matsuda

Augmented City 3D

Keiichi Matsuda (BSc. MArch) is a designer and film-maker. His research examines the implications of emerging technologies for human perception and the built environment. Keiichi is interested in the dissolving boundaries between virtual and physical, working with video, architecture, and interactive media to propose new perspectives on the city. He has exhibited his work internationally, from London's V&A Museum to the Art Institute of Chicago, the New York MoMA, and Shanghai EXPO. He has been published extensively in print and online, and has won awards for design, drawing, speaking, and film-making. Keiichi runs a small future-facing design studio in London, working at the intersection of technology, media, and architecture.

01

Your films portray a future highly integrated with and dependent upon technology. How do you see these representations contributing to a discourse of speculation around technological implementation within society? Do you see your interpretation as an aspirational future or a logical projection?

Visions of the future tend to get lumped into one of two categories; the utopian vision or the cautionary tale. I try not to work to either of these, though; both seem wilfully ignorant, conveniently airbrushing out anything that doesn't fit into their world-views. My methodology is rooted in the school of critical design, using the process of design as a critical tool to explore, speculate, and propose alternative futures. This means trying to draw out the potential of the technologies, and to understand the possible consequences, both positive and negative. It's not all projection though. As a designer, writer or film-maker, you're always making choices about what you want to draw attention to, how you present things.

My motive is a little more complex. I think we're already wrapped up in the world you describe: highly integrated with and dependent upon technology. It often feels as if our future is thrust upon us, or smuggled through in the fine print of the terms and conditions. In many ways the films just reflect our lives today, so by citing the film in an alternative world, we're able to gain the necessary distance to gain a bit of objectivity. We don't have to buy into the techno-utopian visions offered up to us by the tech giants; we can decide the future for ourselves. For me, the first step towards that is to encourage critical thought and debate about the kind of future we want.

02

The depiction of landscape or environment in your films is often generic and reminiscent of common landscapes across North American cities. The layers of augmented reality become a canvas

to personalize and populate with accoutrements that allow this generic backdrop to have a personal presence. Does this say anything about the future of the physical environment and architecture more specifically?

Yes, there was a conscious decision to focus on the mundane, the generic, the everyday. I became interested in post-modernism and semiotics in the context of consumer technology, so a lot of the design work tries to understand the relationship between the physical and virtual layers of the city. I see the functions of architecture splitting between these layers; on one side you have the physical infrastructure that deals with the very basic requirements of safety, shelter, servicing, and transportation. That can be mundane, impersonal. On the other side, the expansion into the virtual realm vastly expands the possibilities of creating not just interesting forms and decoration, but also animation, responsiveness, interactivity, and settings for events. I still classify it as architecture, as it is fundamentally concerned with peoples' interaction with space; but it's maybe closer to game design or film-making than how we think of architecture today. In the *Augmented City*, the physical environment is relegated to providing a supporting infrastructure to augmented experience. It's Scott Brown/Venturi's "decorated shed" brought into the digital age.

There are a lot of exciting possibilities for the design of spaces within this paradigm. One of these is an idea that I call "subjective space"; the idea is that with customization, every person will perceive their environment in a different way. It's almost the opposite of the homogenized, generic physical spaces; the virtual environments are hyper-local, personalized, and potentially constructed by the local community. We will have to stop thinking about cities as stable, static edifices, and recognize them as dynamic, liquid systems that appear unique to every inhabitant.

03

Do you see *Augmented City 3D* and *Domestic Robocop* as a form of science fiction? Are your films about narrative and storytelling, or about design exploration . . . or somewhere in between? More generally, does science fiction become a provocateur that enables or disables future technologies?

I'm quite lucky in a way; there aren't really any stylistic expectations from web video, so no one sits down to watch my films expecting a drama. As a culture, we place a lot of value in storytelling and characterization, but it's never been something that really interests me. Look at *Domestic Robocop*; there's a story, but the story is about making a cup of tea. For me, the narrative is the framework on which the world is draped. A good narrative will reveal that world and draw out its core issues, so the central focus isn't the characters but the dynamics of the worlds they inhabit. In that way, you could say that my films have a similar spirit to many works of science/speculative fiction.

There's an idea of a golden age, where science and technology inspired science fiction writers, who in turn wrote and directed works of fiction that inspired science and technology. It's a beautiful loop as it involves culture and the arts in the production of the future. But that system seems to have disappeared, to be replaced with something much less deliberate. Now, start-ups in incubators around the world create thousands of systems, apps, and models, and we vote with our clicks. It's a race to fill every niche, almost irrespective of actual need. This system relies on the free market to guide it, and undermines our confidence that the future is something that can be consciously designed. I disagree with this viewpoint though. I think it's important that we take responsibility for the future. If we don't like the direction where things are going, we need to understand that it's up to us to change it.

04

The hybridization of the physical and virtual is becoming more and more common. What are the inherent strengths and weaknesses to this merger as we slowly integrate the richness of the physical world with richness of the virtual world?

When people catch on to an emergent idea, there is always a frenzy round trying to bend that idea to every conceivable use. Then in time, we come to recognize what works and what doesn't. At the moment, people are trying to imagine how this merging of physical and virtual could impact various fields, which has given rise to many new areas of interest; social media and e-commerce in the last decade, and now the quantified self, the virtual mapping of the physical world, the internet of things, self-driving cars, just to name a few examples.

It's not a surprise to see that many of these applications promise convenience. More telling than that, though, is the promise of efficiency—or more precisely control; the ability to quantify, record, sort, tag, list, and categorize life. On the other side, the purely virtual experience is improved by incorporating the tactile, the visceral, gaining a new relationship with the human body. The biggest positive element for me is the potential for entirely new forms of artistic expression to emerge. At one point, we will stop thinking about the physical and virtual as separate realities, and see everything as one continuum. Until then, we are still discovering the possibilities.

The weaknesses are subjective. Ubiquitous augmentation means ubiquitous media. It's impossible to overestimate how powerful media will become in this world. It will fuse so completely with our physical reality that the word "media" itself will probably become redundant and disappear. This new reality will stimulate hype, surveillance, and cognitive offloading, while destroying privacy and objectivity. These are weaknesses to you and me, but could be great strengths to corporations, governments, and other interested parties.

05

The notion of interface or heads up display connects us directly to phenomena in the environment. Through real-time feedback, augmentation, and ancillary data streams we gain a fundamentally different image of the world we occupy with these rich quantities of data at our disposal. Do you see this connection with the environment as a way to enhance our understanding of phenomena as we begin to see actions play out in real-time?

Yes, the monitoring and simultaneous visualization of phenomena that has been intangible until now could certainly impact on our relationship to our environment. Live readouts on home appliance electricity usage could encourage us to be more energy efficient. Personalized displays of nutritional data could persuade us to eat more healthily. And of course, quantification and the production of data will always be valuable to industry, total augmentation being a natural extension of existing workflows. Of course there's always a dark side too; the SketchFactor app (and the Ghetto Tracker app before it) help presumably rich people avoid presumably poor city neighborhoods, and that's without even touching on the hugely problematic issue of who owns and has access to all of this data.

Augmented Reality and other technologies in this paradigm have been framed as forces that will deliver greater control, efficiency, and productivity. I'm sure that they will provide all these things and more, but I can't help feel a bit disappointed by the lack of ambition in these goals. Productivity and efficiency are great selling points for businesses, but I question their intrinsic value to our everyday experience of the world. As human beings, it seems to be other things that really give our lives meaning. I think that this technology has the power to go far beyond, and we should be aiming much higher.

06

You use the terms "Electronomad" and "soft occupation" to describe ways of engaging space through augmented realities. Can you expand on these two concepts?

Electronomad is a term that I borrowed from the architect William J Mitchell. It refers to a mode of existence, where a worker can be liberated from their physical surroundings through the use of mobile devices. Wireless infrastructure allows the Electronomad to use any place as a workplace, or—to extend the idea—as a setting for entertainment or social engagements. We're taught Mies' famous maxim "form follows function" in architecture school, but form and function are starting to become unstuck in the presence of personal mobile devices. Now a coffee shop becomes an office, a commuter train becomes a games arcade, and you can read your children a bedtime story from the hospital. This is what I'm referring to when talking about soft occupation. It's a situation in which the power to define programme of a space is partially returned to the user.

A lot of terminology we use to talk about space becomes disrupted in the context of augmented reality. The lines that were drawn between home and work, public and private, even masculine and feminine, seem increasingly hard to pin down. I don't generally like jargon, but a lot of the conversations I have around hyper-reality genuinely seem to require new or appropriated vocabulary.

07

Are there specific technologies you see evolving that will drastically change the way we interact with augmented realities?

The main challenge with augmented reality is the display (if you can call it a display). It's not just a challenge of being able to convincingly overlay the graphics, but also about making the whole package small and attractive enough to make it desirable. There are several different methods being developed in labs around the world, which prioritize different types of experience, but are starting to look promising. In terms of controlling these systems, we've probably got a lot of what we need commercially available right now, from motion controllers, to LIDAR, to the improving state of voice recognition.

This is quite a limited view of AR though. Defined more generally, augmented reality can describe any situation where *soft* information is experienced against the context of the physical world. This could include smart surfaces, tangible computing, wearables, the internet of things, and many other emerging technologies. You might not see those things as AR, but a lot of the design challenges and social implications overlap. The physical nature of our interactions may change based on whichever technologies are adopted fastest, but there are core principles underpinning them all. That's what motivates me to keep producing work. Hyper-reality is coming in one form or another, and it's up to us to design ourselves a future that we want to live in.

NOTES

1 J. Meejin Yoon and Eric Höweler, *Expanded practice: Höweler + Yoon Architecture/My Studio* (New York: Papress, 2009), 133.

2 Ibid.

3 Ibid.

4 Dunne, *Hertzian Tales,* 107 (see chap. 1, n. 18).

5 Shannon Mattern, "Methodolatry and the Art of Measure," *Places Journal,* (November 2013). Accessed 13 Feb 2015. <https://placesjournal.org/article/methodolatry-and-the-art-of-measure/>

6 Anne Galloway, "Intimations of Everyday Life: Ubiquitous computing and the city," *Cultural Studies* Vol. 18, no. 2/3 March/May 2004): 390.

7 Jane Amidon, "Big Nature," In Design Ecologies: *Essays on the Nature of Design*, edited by Blostein, B. & Tilder. New York: Princeton Architectural Press, 2010.

8 Henrik Ernstson, Sander E. van der Leeuw, Charles L. Redman, Douglas J. Meffert, George Davis, Christine Alfsen, Thomas Elmqvist. "Urban Transitions: On Urban Resilience and Human-Dominated Ecosystems." *AMBIO* 39, (2010): 531–545

9 Ibid.

10 Ibid.

11 "Average Happiness for Twitter," http://hedonometer.org/index.html.

12 We Feel Fine mission statement, http://wefeelfine.org/mission.html.

13 Portions of text adapted from the design statement developed by Bradley Cantrell and Carl Koepke

14 Höweler and Yoon, *Expanded Practice*, 144.

15 See, project description for "Audio Grove" in Lucy Bullivant, *Responsive Environments*, 35 (see chap. 1, n. 20).

16 Anthony Dunne termed "hertzian landscape" in Dunne, *Hertzian Tales.*

17 Haque, quoted by Bullivant in *Responsive Environments*, 63.

18 Haque, "The architectural relevance of Gordon Pask," 55.

19 Bullivant, *Responsive Environments*, 63.

20 Dunne, *Hertzian Tales*, 111.

21 Haque quoted by Bullivant, *Responsive Environments*, 63.

22 Dunne, Hertzian Tales, 121.

23 Natalie Jereminjenko, "Introduction" The xClinic Environmental Health Clinic, NYU. http://environmentalhealthclinic.net/environmental-health-clinic.

24 Woebken interview in Manaugh, *Landscape Futures,* 90 (see chap. 4, n. 10).

25 Ibid.

26 Ibid.

27 David Benjamin and Soo-In Yang Interview in Manaugh, *Landscape Futures,* 99.

28 Claudia Pasquero and Marco Poletto, "HORTUS," ecoLogic Studio (blog), January 13, 2012, http://ecologicstudio.com/v2/project.php?idcat=3&idsubcat=59&idproj=115.

29 Claudia Pasquero and Marco Poletto, "About," ecoLogic Studio (blog). http://ecologicstudio.com/v2/about.php?mt=1.

30 Claudia Pasquero and Marco Poletto, "HORTUS."

31 Gilles Clément, *Manifeste du Tiers paysage,* (Montreuil: Sujet-Objet, 2004).

32 Claudia Pasquero and Marco Poletto, "HORTUS."

33 Lucy Bullivant, "Algae Farm," Domus (blog), September 16, 2011, http://domusweb.it/en/architecture/2011/09/16/algae-farm.html.

34 Matthew Gandy, "Cyborg Urbanization: Complexity and Monstrosity in the Contemporary City," *International Journal of Urban and Regional Research* 29, no. 1 (2005), 26–49. doi:10.1111/j.1468-2427.2005.00568.x.

35 Keiichi Matsuda "Domesti/City: The Dislocated Home in Augmented Space," Master of Architecture Thesis, University of College London, Bartlett, 2010.

36 Ibid.

AMBIENT

08

Ambient considers interventions and installations presenting information abstractly, precisely, and repetitively over longer periods of time, requiring a learned relationship for interpretation. Humans rely on a relationship with phenomena to make judgments about the environment based on past experience. Typical descriptions of ambient phenomena include: quality of light, shifts in wind, temperature, and seasonal change. These environmental qualities provide a datum for learning about the environment through ambient cues, forming deeper relationships with context. Often these cues are established by attaching ambient phenomena to cyclical or fundamental dynamic processes. Through experience we learn to interpret these contextual cues such as the smell and feel of water laden air as the harbinger of rain or the rush of cold air to signal a weather front on the horizon. Through technology we develop similar relationships: the chime of the bell tower to signal time or the purposeful dispersion of scent to lure customers into a fast food restaurant.[1] Each event requires a learned relationship that slowly places the processing of that event in the background. There are indicators, in which a sense or multiple senses announce specific shifts or alterations in relation to developed associations. Associations will tie indicators to environmental conditions in the form of classical conditioning. In this sense, environmental or contextual literacy is developed through comparisons requiring repetition and a known datum.

The definition of "ambient" is "an encompassing atmosphere," "existing or present on all sides," and "encompassing."[2] In many ways, "ambient" is a metaphor for most continuous environmental processes such as ecological and biological systems exhibiting a perceptible vitality or fecundity through responses to growth and decay over time. The term "ambient" has been connected to the concept of integrating computing into environments and futures associated with ubiquitous computing, hybrid realities, augmented realities, and sentient environments for some time. Rather than focusing solely on the object and how it interacts with the environment, the term "intelligent ambience" speaks to "distributing that intelligence throughout an environment."[3] Anthony Dunne quotes Peter Weibel:

> Machine intelligence will serve to make the environment more efficient and more intelligent so that it will be able to respond more dynamically and interactively to human beings. The realisation of the concepts of computer aided design and virtual reality will thus be followed by the realisation of computer aided environments and intelligent, interactive, real surroundings. The latter will be referred to as intelligent ambience—an environment based on machine intelligence.[4]

The clock tower is an important icon of ambient information. Before the advent of the wristwatch, the clock tower served an important purpose as the source of a synchronized time piece. These structures started as bell towers that used the ringing of bells to signal church services or times of prayer. Over time clock faces accompanied the bells, allowing a visual indication of a more precise time. This ambient information was viewable with a glance or as a background noise; as information, it was also tied to environmental cues and could be interpreted in tandem with the sun in the sky. As personal timepieces became ubiquitous, the need for the clock tower waned. Instead, individuals kept time by syncing their watches to one another or to central broadcasts. Today ambient temporal information is removed from environmental cues and is placed solely into official timepieces that govern the operation of time in our devices.

Ambient speaks to habit: the repetition of actions that occupy a peripheral space in the human mind. This is the ability to perform a task with very little processing, a form of autonomy.[5] Our lives consist ". . . of a multiplicity of rhythms. Everyday life thus entails a range of flows, each with their own proper time."[6] The background processing of phenomena creates the potential to perform a range of activities and, rather than focusing on a single task, ambience processes multiple phenomena. These are stable relationships that only form after learned experience. The work of David Rose at Ambient Devices, founded in 2001 to commercialize patent-pending technologies developed at The Media Lab, involves building devices that inhabit ambient environs by developing habits that require very little processing to interpret. The *Ambient Orb* serves to visualize personalized phenomena such as temperature, stock prices, conditions of ski slopes, or even an indicator of waves at your favorite surf spot. The device is a translucent orb that displays a single stream of data, lit by RGB LEDs allowing it to display a range of colors.[7] *The Ambient Orb* is meant to inhabit a table top, kitchen counter, or other open space and synced with a data stream. Based on the synced data, the orb will illuminate with a range of colors: for instance, if the weather forecast calls for 75 degree weather the orb will have a light yellow hue indicating a warm day, if the weather is cooler it might display a light blue color. Over time the user can merely glance at the orb and rely on the information to plan the upcoming day. The nuanced color displays give the consistent user an ambient understanding of weather—or any other data stream that may be attached to the device.

Through learned behavior it is possible to develop contrary relationships to a designer's intent. For example, in cities the ambient cacophony of car alarms serves to create a backdrop of complacency rather than concern that an automobile is being stolen. The usual reaction to a car alarm is that it is a nuisance, particularly after a string of consecutive false alarms. This relationship with the car alarm curates an ambience that can be acted upon or ignored. An urban monitoring project by the Oakland, California police department, ShotSpotter, is a "network of microphones that detect gunfire in most parts of East and West Oakland."[8] The network uses an array of microphones through the city to not only detect gunshots but also co-locate the source. The system has the ability to detect a range of ambient information, but has been specifically tuned to the frequency and duration of gunshots.

With computation surrounding humanity, the act of computing itself is ambient. The implied connections are delicate and fragile linkages that become robust through continued interaction producing learned states. "The most potentially interesting, challenging, and profound change implied by the ubiquitous computing era is a focus on *calm*," a serenity that provokes a reliance on the manipulation of salient connections—a continual reprioritization of data in an ambience of connectivity, in which ". . . calmness is a fundamental challenge."[9] How do we operate in a space of habits, routines, and rhythms? What are the methods of manipulation for designers? In an environment of plurality—a background of information—it becomes necessary to develop means to intersect, interrupt, and align responses. This suggests that ambient projects operate in real-time—processing a current state and responding accordingly—as a form of technological breathing, inhaling the environment and processing it. A continued barrage of technology displaces discrete connections with environmental phenomena. Rather than disregarding technological devices as a source of distraction, is it possible to build more meaningful connections in real-time?

Bloom

Los Angeles, California, 2011

Doris Sung, Ingalill Wahlroos-Ritter, Matthew Melnyk, dO|Su Studio Architecture

Collaborators: Doris Kim Sung (Principal, DOSU Studio Architecture, and Assistant Professor, USC), Ingalill Wahlroos-Ritter, Glass Consultant (Principal, WROAD and Chair, Woodbury University), Matthew Melnyk, Structural Engineer (Principal, Nous Engineering)

Design Team: Dylan Wood (Project Co-ordinator), Kristi Butterworth, Ali Chen, Renata Ganis, Derek Greene, Julia Michalski, Sayo Morinaga, Evan Shieh

Construction Team: Dylan Wood, Garrett Helm, Derek Greene, Kelly Wong (Core Contributors), Manuel Alcala, Eric Arm, Lily Bakhshi, Amr Basuony, Olivia Burke, Kristi Butterworth, Jesus Cabildo, Shu Cai, Ali Chen, Taylor Cornelson, Erin Cuevas, Matt Evans, Chris Flynn, Renata Ganis, Bryn Garrett, Ana Gharakh, Oliver Hess, David Hoffman, Alice Hovsepian, Casey Hughes, Ross Jeffries, Justin Kang, Syd Kato, Andrew Kim, Glen Kinoshita, Ingrid Lao, Jennifer MacLeod, Max Miller, Mark Montiel, Laura Ng, Robbie Nock, Raynald Pelletier, Elizabeth Perikli, Nelly Paz, Evan Shieh Hector Solis, Raven Weng, Leon Wood, Tyler Zalmanzig

Funding was generously provided by: AIA Upjohn Research Initiative, Arnold W. Brunner Award, Graham Foundation Grant, USC Advancing Scholarship in the Humanities and Social Sciences Program, USC Undergraduate Research Associates Program, Woodbury Faculty Development Grant, and in-kind donations from Engineered Materials Solutions

Scope: Temporary Installation at Materials & Application Gallery, Los Angeles, CA

Technology: Thermobimetals, Parametric Modeling and Analysis, Digital Fabrication

---

Bloom is a materially responsive sun-tracking device constructed from thousands of smart thermobimetal tiles actuated to bend and curl in response to ambient temperature and direct sun exposure. Doris Sung, the principal investigator, is part of a growing field in digital and parametric architecture investigating the potentials of digitally fabricated metals. Sung's main research has been with "smart" thermobimetals. These are normally used in common thermostats to detect room temperatures, but never before at an architectural scale. The thermobimetal material is a thin sheet of metal composed of two alloy layers, each with different sensitivities to heat, sealed together. When the thermobimetal is heated, one of the metals expands faster than the other, causing the sheet to curl in response with no methods of control or energy input. Capitalizing on the material's attributes, Sung has researched and fabricated complex architectural forms by cutting, tiling, and weaving the thermobimetal sheets to strategically morph under specific temperature conditions.

Figure 08.02 Thermobimetal tiles. Bloom, *dO|Su Studio Architecture Los Angeles, California, 2011*

Figure 08.03
Long exposure. Bloom, *dO|Su Studio Architecture Los Angeles, California, 2011*

Sung is currently prototyping methods for responsive and zero energy temperature control in warmer climates with responsive skins, brise soleils, and shade structures; and in cooler climates using the material to devise heat activated self-assembly systems.[10] *Bloom* is the result of research geared towards the use of thermobimetals for responsive building skins that passively regulate building temperatures. It was the first project to show the material performance at a large scale within an environment. Sited at the Materials & Application Gallery in Los Angeles, the structure was outdoors and between two buildings, subject to a range of thermal conditions by fluctuating sun exposure and ambient temperatures.

Bloom was designed with dual functions: to showcase the use of thermobimetals as a sun shading device, and as a prototype for temperature regulating building skins. Composed of thousands of individual bimetal tiles assembled into 414 paraboloid-shaped stacked panels, *Bloom* is an independent structure, "dependent on the overall geometry and combination of materials to provide comprehensive stability."[11] The thermobimetals were embedded with logic to close and prevent heat from entering a space in high temperatures as well as open for ventilation to allow heat to escape through apertures.

As the metals respond to temperature changes, the structure visibly indexes: time and temperature through the overall form, the apertures within the panels, and the amount of shade cast by the structure. To achieve such complex and nuanced material performance, the designers used Grasshopper and parametric modeling tools throughout the entire design process and into fabrication: "we use them

Figure 08.04
Bloom, *dO|Su Studio Architecture Los Angeles, California, 2011*

to form-find; we use them to generate the fabrication files; we use them to analyze the structures and project the performance; and we use them to test for post-occupancy performance."[12] The use of computational tools allowed the designers to predict the heat intensity on the surfaces by running solar analysis on digital 3D models. Incredibly site-specific temperature changes actuate localized responses to individual thermobimetals resulting in slight ambient fluctuations to individual tiles that aggregate into overall structural movement and shifting light conditions across the space.

Bloom is emblematic of the goals of responsive architecture to create more sustainable and environmentally responsive building systems.[13] The ability to embed materials with specific logics tuned to perform in specific environmental conditions is of course more energy efficient for responsivity, but also has the potential to be responsive in a way that cannot be computed. As we learn more about material relations within complex systems—especially novel or emergent ecological conditions—there is substantial potential to indicate specific conditions with more evolved iterations to cue embedded material responsivity.

Figure 08.05 Bloom, *dO|Su Studio Architecture Los Angeles, California, 2011*

Figure 08.06 *Fins with Nitinol wire, Reef, Rob Ley/Urbana and Joshua G. Stein/Radical Craft, Storefront for Art and Architecture gallery, 2009. Photograph by Alan Tansey*

Reef

New York City, 2009

Rob Ley and Joshua G. Stein

Collaborators: Rob Ley/Urbana, Joshua G. Stein/Radical Craft

Interaction Concept and Development: Active Matter, Design Team: Timothy Francis, Jonathan Wimmel, Elana Pappoff. Fabrication Assistance: Peter Welch, Daniela Morales, Travis Schlink, Joshua Mun, Darius Woo, Lisa Hollywood, Rafael Rocha, Yohannes Baynes, and Phillip Ramirez

Support: *Reef* was supported by grants from the AIA Upjohn Research Initiative Program, the Graham Foundation for Advanced Studies in the Fine Arts, AIA Knowledge Grant, and IDEC Special Projects Grant. This project was made possible by the generous support and assistance of Dynalloy Inc.

Scope: Temporary responsive installation for the Storefront for Art and Architecture gallery

Technology: Motion control software and Custom Electronics: Pylon Technical

---

Reminiscent of a curious new species, *Reef* is an installation of moving fiber reinforced composite scales deformed by shape memory alloys, metals which contract and release in response to temperature change. The actuation of the aggregation of scales across the

Figure 08.07
Reef, Rob Ley/Urbana and Joshua G. Stein/Radical Craft, Storefront for Art and Architecture gallery, 2009. Photograph by Alan Tansey

Figure 08.08
Reef, *Rob Ley/Urbana and Joshua G. Stein/Radical Craft, Storefront for Art and Architecture gallery, 2009. Photograph by Alan Tansey*

3-dimensional surface reference the movement of plants and lower level organisms with the capacity to automatically respond to environmental phenomena, similar to sun flowers or sea anemones.

Reef was designed for and exhibited at the Storefront for Art and Architecture, as a responsive mediator between degrees of private and public space. Evolving over years of research with shape memory alloys as actuation devices, *Reef* represents a catalyzing moment within a larger body of research by Rob Ley and Joshua Stein "to shape the public perception of the built environment through the activation of membrane and partition surfaces."[14] Novel to the field of responsive architecture at the time, *Reef* offered more than interaction with a singular object or 2-dimensional surface by activating and defining space.

Composed of over 900 responsive fins attached to an aluminium frame, *Reef* negotiates physical and social movement between the linear gallery space and the streetscape. The flexing and opening of the fins is cued to movement in the gallery by isolating shadows detected by an RGB camera to activate sections of the surface—offering intriguing participation by revealing new spaces and connections through the responsive apertures. Although the fins were actuated in sections across the surface, each fin and Nitinol wire actuator was individually fabricated and tuned to uniquely respond. The installation was also programmed to move and actuate when there was no sensed phenomena (no one in the gallery) to establish interest and curiosity. These methods of actuation result in ambiguous, almost life-like behaviors, moving organically rather than an established binary reaction.

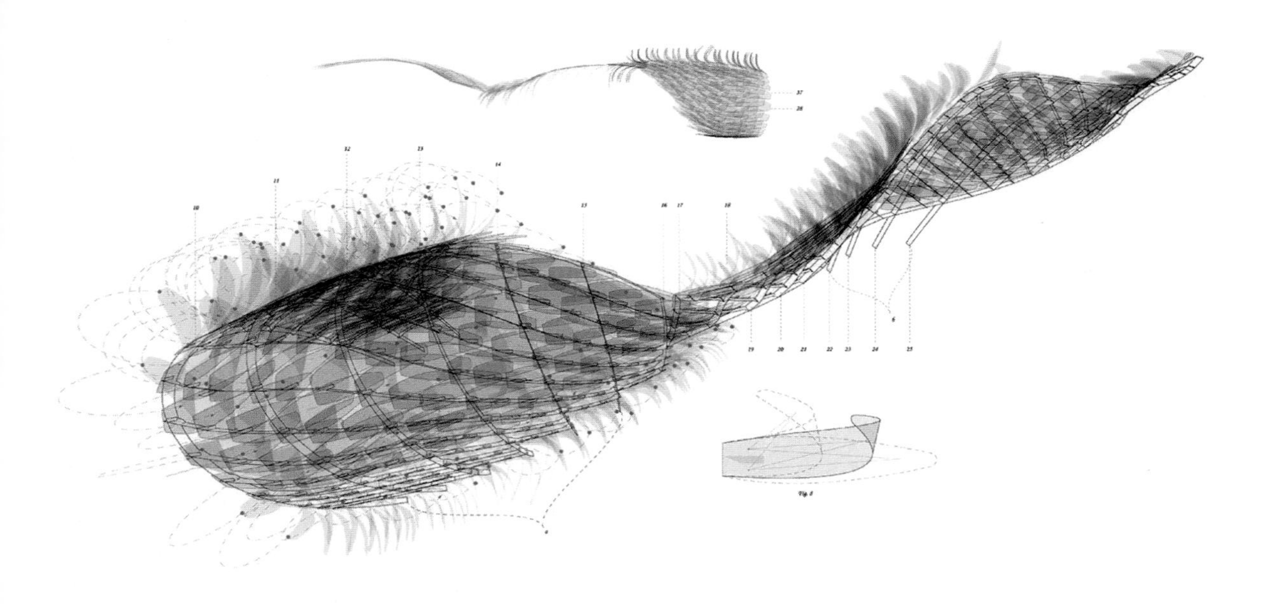

Figure 08.09 *Movement diagram,* Reef, *Rob Ley/Urbana and Joshua G. Stein/Radical Craft, Storefront for Art and Architecture gallery, 2009*

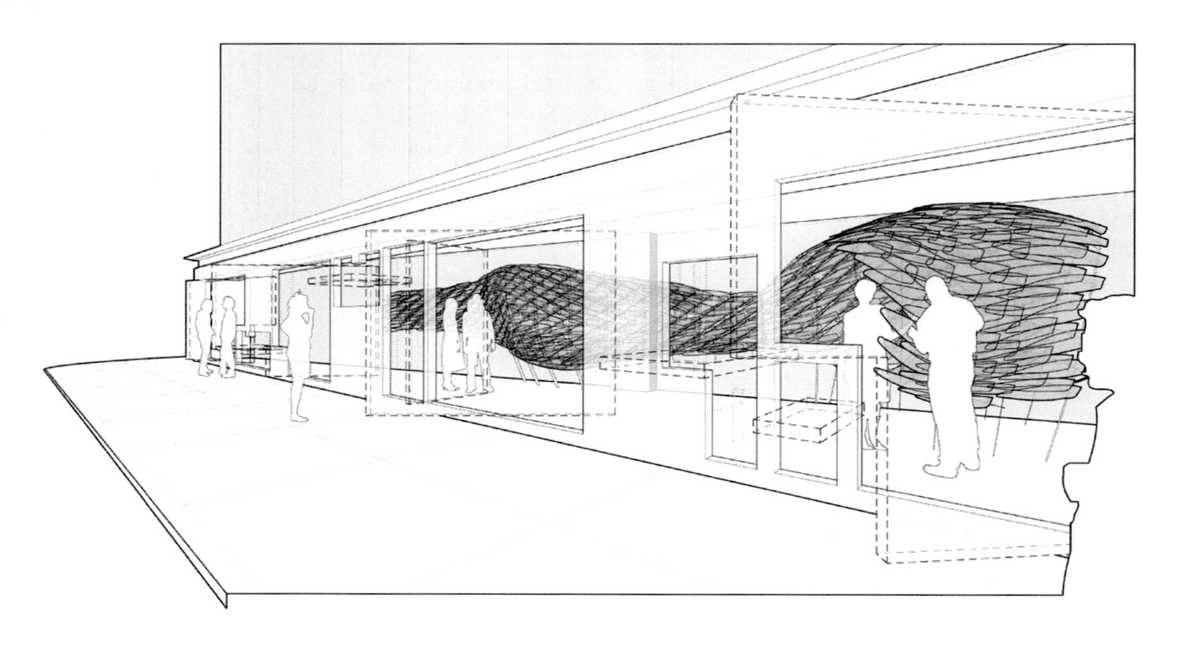

Figure 08.10 *Concept drawing,* Reef, *Rob Ley/Urbana and Joshua G. Stein/Radical Craft, Storefront for Art and Architecture gallery, 2009*

Figure 08.11
Photograph from within the barrel, Reef, Rob Ley/Urbana and Joshua G. Stein/Radical Craft, Storefront for Art and Architecture gallery, 2009. Photograph by Alan Tansey.

Although *Reef* was very much designed within the realm of responsive architecture, as a precedent for responsive landscapes it blurs the boundary between biotic behaviors and abiotic material. Methods of material actuation are particularly important for landscape architecture because of the responsive capabilities tied to the dynamic qualities of landscape materials. Similarly to *Bloom*, embedding an "intelligence" or "technological capacity" into physical materials has the potential to engage with dynamic landscape systems and external flows of material and energy fluctuations. With many responsive installations the process from sensed phenomena to actuation has a binary application and results in a clear translation of either on or off—for example an LED light. With *Reef*, the method for actuation remains the same (supplying a current), but the application of intelligently fabricated materials offers another layer of translation. Through careful computation and material calibration, the installation takes on a life of its own. In *Reef*, the blurring of biotic and abiotic material behaviors elicits empathy[15] or emotional attachment in the viewer. The cultural or ecological significance of landscape elements—particularly aspects of landscape related to performance—are not often perceived as aesthetically beautiful or relevant. Thus, evoking an emotional or relational significance offers new approaches for endearing or engaging with material-based processes.

Interview with Joshua G. Stein

Reef (a collaboration with Rob Ley/Urbana)

Joshua G. Stein is founder of the Los Angeles-based studio Radical Craft and the co-director of the Data Clay Network, a forum for exploring the interplay between digital techniques and ceramic materials. Radical Craft (*www.radical-craft.com*) operates as a laboratory for testing how traditional phenomena (from archaeology to craft) can inflect the production of urban spaces and artifacts, evolving newly grounded approaches to the challenges posed by contemporary virtuality, velocity, and globalization. He has taught at the California College of the Arts, Cornell University, SCI-Arc, and the Milwaukee Institute of Art & Design. He was a 2010–2011 Rome Prize Fellow in Architecture, and is currently Associate Professor of Architecture at Woodbury University.

01

The responsive qualities of *Reef* have been compared to the physical and biological responses of phototrophic plants and invertebrate coral reef species. The growth response of sunflowers to sunlight and the enigmatic movements of sea anemones are actuated by biochemical and physical responses to environmental stimuli. The responses manifest in slow seemingly static ambient movements revealed over time. Similarly, *Reef* operates through the aggregation of subtle material driven responses without an apparent central system of control. How was the design and narrative of *Reef* influenced by similar examples of embedded logics?

We looked to sea anemones and sunflowers because they produced both a design narrative and a model for defining a field condition with a certain amount of unity operating across a set of similar yet discrete elements (like a time lapse video where you see many sunflowers rotate in unison to track the sun's movements). We knew we wanted the fins of *Reef* to move at certain moments in unison and at others to operate in direct response to some kind of outside stimulus, most likely to the movement of people in the gallery. This is actually how we decided on the title *Reef*. In the context of a coral reef, sea anemones exhibit two different types of movements: an ambient rhythmic sway with the ocean's currents, and a more direct and local motion where one or several anemones might react to a fish swimming by. We were interested in designing our installation to operate in both modes.

In terms of regulating the motion of *Reef*, the installation configuration required the input of one RGB camera, establishing a centralized system of control. However, there was actually no way to control all the fins simultaneously because they were split into different channels. The system of actuation was composed of independent sets of approximately 12 fins that moved as a group. We called these groupings "patches." Within these patches, there were material discrepancies due to the fabrication of each fin and the length of the

Nitinol wires that would guarantee individual nuances distinct to each patch, even though every patch is receiving currents controlled by a centralized source. The length of the Nitinol (a shape memory alloy that changes its size according to heat) affects how long it takes to heat up and for the fin to curl up. Ultimately, there was never going to be an option where all fins moved as a unified set because of the physical differences between any module and its neighbor. This was in keeping with the initial design sensibility, although not necessarily pre-planned. We didn't have a desire to make all fins appear identical or completely subservient to a larger system. There's automatically a certain amount of variation in motion that emerges when each fin is conceived and produced as a distinct module.

02

What was your process for prototyping and testing the responsive objects? How did the material palette and material performance shape the overall architectural form?

The project demanded years of research, including a couple years dedicated solely to material prototyping for both the responsive technology and the material components. This started with empirical testing of the Nitinol muscle wire to understand its behavior and to establish the best way to attach the wire to a fin material. When we first started out we didn't know what the relationship would be between a linear actuator and the material component that could occupy 3-dimensional space. At first, we thought this might be something like Chuck Hoberman's toy spheres that mechanically expand and contract. Ultimately, we settled on something that started out as perfectly 2-dimensional and planar, and then morphed to occupy 3-dimensional space as a deformed surface. This strategy aligned nicely with some of our precedent studies—we had found a toy butterfly that would "flap" its wings, actuated by a tiny piece of Nitinol (we contacted the manufacturers and ultimately used the same proprietary material, Flexinol, in the *Reef* installation). It was a nice example of a minimal amount of linear motion in the wire amplified geometrically over the form's surface to produce a larger visual impact. Basically, the butterfly's wings move a significant amount even though the actual motion of the Nitinol is barely perceptible. Employing that strategy proved to be both the best use of our budget (the Nitinol can be quite expensive) and a visually compelling use of the technology that anyone could appreciate regardless of their interest in the technical aspects of the project.

Through this round of material prototyping we tested methods of coupling the Nitinol with other materials that could visibly register the wire's change in length. An earlier version was published in *Interactive Architecture* by Michael Fox and Miles Kemp, depicting a long fin cut lengthwise down the middle with two different Nitinol wires, one running along the top panel and the other along the bottom. Each panel bent in a different direction resulting in a dramatic transformation of a planar fin into something that truly occupied 3-Dimensional space. Our original strategy of arrangement was somewhat like a voxel scheme, gridded out with modules—similar to doors or gates—that, when actuated, functioned like a set of corridors, opening and closing down passages. When we were invited to produce the installation at the Storefront for Art and Architecture in New York, it became clear this scheme wasn't appropriate proportionally or dimensionally for that extremely narrow site. We instead decided to fill the gallery with one large vaulted surface articulated with smaller fins that could exhibit motion across its length, filtering the views between the microspaces within the gallery rather than attempting to open or close down circulation paths with larger fins. When we scaled the fins down, we simplified each module to have one actuator instead of two.

At this point, there was still a lot of prototyping to determine the appropriate 'fin' material and attachment method. We tried all different sorts of polymers/plastics, and in the end we settled on a material used for circuit board backing called G-10. The Nitinol only operates by contraction in one direction. Once you run a current through the wire, it gets hot, contracts, and doesn't return to its original shape immediately unless it is coupled with a spring-loaded material to pull it back after the current is off. If you were to hook this up to a piece of panel material that wasn't rigid enough to pull it back into its original shape, it would simply stay curled until you flattened it out by hand. We needed a material that would return to its original shape after we turned off the current but that was also lightweight. We tested polypropylene, polycarbonate, all the polys—some of these materials would have been easier to fabricate with the laser cutter—but in the end, because we essentially ended up with a fiberglass resin that we couldn't laser cut, we had to use a CNC router for everything which was messier and took longer.

Once we figured out the technical application of the Nitinol, there was another set of material testing to calibrate the software to produce the physical motion for each patch. To actuate the fins the power needs to ramp up very quickly, then ramp down quickly, sustain for a few seconds, and then drop off slowly, allowing the fin to pull the wire back into shape. We spent a lot of time with our consultant, Ben Dean of Pylon Technical, who was working with Max MXP software to properly calibrate that power curve for each patch of fins depending on their size and orientation. Those on the top of the aluminum vault structure were obviously affected by gravity differently from those on the side. We had to pretension each fin to have a camber so that it would contract in the correct direction, and we could also pretension a bit more to overcome gravity for the bigger fins or those on the top of the vault. Tuning that power curve was a delicate balance because we wanted the movement to feel more biological than mechanical, and ramping up the power too quickly to overcome gravity could also produce a jerkiness—which could ruin the overall effect.

We used four Arduino boards to control the motion of the fins and one RGB camera to track the movement and proximity of the visitors. The motion of the visitors' shadows was the easiest way to track human movement within the gallery without distractions; although people might have different colors of clothing, everyone's shadow is the same color. If you walked towards the piece, the fins in closest proximity would be triggered to curl towards you. As you continue moving along the piece, the closest fins would curl accordingly, tracking your movement trajectory across the gallery.

03

The ambient movements of *Reef* inspire wonder in the viewer, remaining disconnected relative to most interactive installations with discrete one-to-one responses between the viewer and media. There is a spatial ambiguity between the users, the phenomena, and the physical response of the piece. How do you define this example of interaction?

It was important to let the public know that the piece was tracking their motion, which required a certain amount of one-to-one responses that allowed you as a viewer to perceive that the piece *knew* you were there. But there were also moments when there was no one in the gallery, and it was necessary for *Reef* to move of its own accord to *invite* people into the space. Even in the second iteration at the Taubman Museum in Roanoke, where *Reef* was positioned at the end of a much longer gallery, we needed to know that, even if nobody was nearby, it would still produce occasional movement, as if the ocean currents had just shifted. We thought of it almost like a jukebox in a bar—even if no one puts any money in it it's still going to play on its own every once in a while. Then when you would put your money in, it would override whatever random pattern it would have of its own. On the programming end of things, this meant there were basically two sets of instructions. We had a set of pre-programmed patterns that would essentially be on shuffle that were relatively subtle but still visible enough to draw someone into the gallery. Then whenever someone would enter into its sphere of perception, it would override whatever pattern it had been on in order to respond directly to the visitor.

We hoped the motion would have a logic of its own but still respond to outsiders. We were trying to produce some form of empathy or even projected personality onto the piece so that the public would believe it had more intelligence—perhaps even intentions or emotions—than it actually did. Our original interest in the technology was tied to this potential. At the time Sony had just come out with the robot dog Aibo and not long after that Honda came out with the Asimo robot. We were interested in how these responsive 'toys' either felt like a machine that simply follows commands or instead behaved like something that could have more of a personality, more like a pet or even a friend. At the time, those robots weren't programmed to have personality, but there were still certain moments where you could start to project a personality onto them. We wanted *Reef* to be more than simply responsive. We wanted it to elicit or accept some form of projected personality.

04

Once *Reef* was installed, what specific interactions or behaviors did you observe? Were there any trends or unanticipated reactions?

Surprisingly enough, for the most part we predicted the way people would react to *Reef* responding to them. What we anticipated less was how important the role of association would be to the public. Just the name *Reef* changed the way people perceived it. It didn't necessarily change the way we thought of it since we had a desire to produce this kind of interaction long before the analogy of *Reef* existed to us.

When we were installing at Storefront, we kept the pivoting doors/windows open and everyone would walk by and ask us what it was. Was it a dragon? A bird? The wing of a bird? For us, we wanted people to approach this without knowing exactly what type of species it was. We had been looking at examples of eighteenth and nineteenth century drawings during the rush to understand newly discovered species, like the work of Ernst Haeckel. We hoped for the public to approach *Reef* in the same way—with curiosity and potentially even trepidation. Titling the installation *Reef* gave it an association without necessarily dictating exactly whether it was plant, animal, environment, or object. It's not a giant fish, it's not a giant bird. However, it is biomorphic, or at least biokinetic, and it's not just about abstract form or abstract motion. It alludes to a specific kind of movement while remaining ambiguous as to whether it might refer to the scale of a reef or to the scale of an organism inhabiting the reef.

05

What is the significance of your method within the application of responsive technologies? How do you contextualize *Reef* within the larger body of responsive architecture and installations?

I haven't necessarily tracked its influences on the larger body of responsive architecture. It's easier for me to talk about *Reef* in terms of what came before it rather than what came after it. At the time, within the field of responsive architecture, everyone was trying to show a direct one-to-one relationship between stimulus and response; less concerned with the associations and more concerned with direct registration. Many did amazing things in terms of actuators and responsivity, but the work tended to be 2-D—content to be on a wall. We set out very clearly to *not* create a demonstration of technology. We looked at interactive pieces of art concerned with the novelty of technology and asked the question: what can we offer as architects? Rather than following the 2-Dimensional trend at that time, we set out to create one of the first installations that was responsive as 3-D form and space.

This was a difficult challenge given the constraints and the acute angle of the Storefront space where there wasn't room to produce much beyond the scale of a surface, so we opted to run with that as a strategy. Even though we were pushing against the 2-dimensional version of responsivity, we decided to let one side operate as a billboard to call passers-by into the space. We extended the 3-dimensionality of the surface by making one simple move to roll the surface over to form an inhabitable barrel vault. This move divided the gallery into several spaces: in between the surface and wall, underneath the barrel vault, between the installation and the iconic exterior operable wall of the gallery, and the public sidewalk space outside of the gallery. With this division of space, the installation became a social mediator between different degrees of public and private spaces. Clearly, outside of the gallery, it is the most public and as you move deeper into the installation it continues to remove you from the streetscape. We also worked very carefully to structure and align the movement of the fins to create apertures along sight lines, opening up visual connections between the different micro-spaces that were parceled off within the gallery space.

06

Reef is an example of embedding life-like qualities exemplifying natural phenomena in objects with behaviors attributed to biological and intelligent capabilities. How does this alter the perception of material performance?

We were never interested in showcasing the materials or the specifics of the technology. With the Nitinol wiring, we carefully designed the visual patterning of how the exposed wiring would weave through the structure and, although we decided not to hide anything, we weren't explaining exactly what was happening either. At almost every step we opted for *performance* in the theatrical or dramatic sense of the word instead of the idea of "performance criteria" often discussed in architecture. For example, we approached scripting as you would in theater rather than scripting in terms of coding. The phrase "material performance" is difficult; *Reef* was innovative in terms of material and technology, but we weren't trying to rationally demonstrate the material technology to the public.

However, certainly in terms of what we asked of the discipline, it was probably one of the first projects to use Nitinol wire at that scale, aside from Philip Beesley's *Hylozoic Ground*. In terms of the *behavior* of the installation, we were trying to maintain the biomorphic and biokinetic allegory at every level. We wanted the potential of this material innovation to be understood in terms of companionship rather than utility—you might approach it more like a pet than a new stereo system or gadget. We never explained the technology, and yet nothing was off-limits to the public. Everyone touched it. Yes, these were high-tech materials—one coming from aerospace and biomedical engineering and the other coming from Silicon Valley—but that novelty wasn't the point of the project. Instead, the viewers were asked to have or project some kind of emotional attachment to a technology that they weren't necessarily meant to understand rationally.

07

What role do you see material performance playing in the future of responsive design?

We are not interested in technology as a goal in and of itself. The desire to produce more and more technological novelty is not why we engaged in this discourse and neither of us pursue responsive technologies as the conceptual driver in our independent work. Rather, when I think about technology as applied to some of my current work in ceramics, it's more about tapping into using the associations we already hold with that material and the techniques involved and trying to reveal or coax out another reading. That was the case for *Reef* as well.

Given current momentum and the nature of the people who generally engage in technology research, the future of technology and of responsive design will probably be performance-driven in the technical criteria sense of the word *performance*. As designers, it was appropriate for us that *Reef* did not necessarily derail that trajectory, but instead ask why architects get in this game when there are clearly engineers and scientists who have been doing this for longer and who have access to better funding. I could use the word poetics to describe what we can offer as designers, but perhaps it's really more about perception and coding material and technology with associations different from that of pure utility.

Figure 08.12 METAfollies, Marco Poletto and Claudia Pasquero, ecoLogic Studio, FRAC Center in Orleans, France, 2013

METAfollies

Orleans, France, 2013

ecoLogic Studio

Collaborators: Claudia Pasquero and Marco Poletto

Andrea Bugli, Philippos Philippidis, Mirco Bianchini, Fabrizio Ceci, Phil Cho, George Dimitrakolous, Manuele Gaioni, Giorgio Badalacchi, Antonio Mularoni, Sara Fernandez, Daniele Borraccino, Paul Serizay, Maria Rojas, Anthi Valavani

Scope: Permanent Interactive Installation at the FRAC (Regional Contemporary Art Fund) Center in Orleans, France

Technology: Piezo Buzzer, Microprocessor Arduino, Proximity sensors

---

With the design of *METAfollies*, Claudia Pasquero and Marco Poletto, founders of ecoLogic Studio, have woven together concepts surrounding the material conditions of contemporary urban landscapes—particularly the landscape of "urban trash," expanded from the concept of garbage to include an assemblage of "products, landscapes, media content, attitudes and lifestyles." Through a feedback of "meta-language" the *METAfollies* ask contemporary society to reconceptualize and evolve our role with "urban trashing" through radical ecologic thinking and material activism. Commissioned for the permanent collection at the FRAC Center in Orleans, *METAfolly* is a "sonic environment" digitally fabricated from urban trash and coupled with responsive technologies. Humming similar to the sound of swarming crickets emerges from alien and organ-like pods—the "synthetic organism" records movement and plays back sounds between the visitors and the installation as a "real-time meta-conversation" about the physical and material pervasion of urban trash."[16]

METAfolly was inspired by historical concepts of the architectural folly originating from the Romantic English Landscape as a method for playfully engaging "nature." ecoLogic studio was particularly interested in the "grotto" for its "immersive environment" and ability to "[fake] the spatial effects of a natural cave." Likewise, the *METAfolly* develops a new naturalness from lifeless materials and technologies—and from the synthetic combination of localized recordings and interactions—simulates complexity. The project was developed over a month long workshop in which a team of young architects were asked to act as "cyber-craftsmen" to hack low-cost responsive technologies in combination with the materials of urban trash. Working with off-the-shelf parts, cheap Chinese gadgets, and machining recycled plastic panels, the team developed customized assemblages of found and recycled objects—termed by Pasquero and Poletto as "slow-prototyping" in reference to the contemporary need for "slow architecture," where the architecture is able to "simultaneously embody the object, the process and the interface."[17]

Figure 08.13
METAfollies, *Marco Poletto and Claudia Pasquero, ecoLogic Studio, FRAC Center in Orleans, France, 2013*

The physical structure of the *METAfolly* is tied to the placement of responsive technologies to encourage interaction and distribute sound. The overall form provides three unique interactive access points to draw the viewers in, record their proximity, and translate it into sound. The physical system is composed of numerous digitally fabricated and technological components organized by a hierarchy of tiles, clusters, tendrils, and hubs. The tiles tessellate together to form the parametric skin of the structure, the clusters of tiles hold singular "active" tiles connected to "active" tendrils (in which active refers to the presence of technological components), and each tendril hosts a piezo buzzer. There are a total of 300 active tendrils within the *METAfolly*. The tendrils encase wires connecting the piezo buzzers to Arduino microprocessors and proximity sensors at the bases of all six hubs. The proximity sensors are capable of sensing visitors from up to five meters away. Once proximity and movement is sensed, the piezo buzzers are actuated. A system of delays and levels of inertia can be tuned and adjusted in real-time to behaviors of movement to synthetically mimic complex behaviors such as the swarming sounds of crickets: "overall the swarm would always escape you but with ever-changing behaviour and sound patterns."[18] With the limited

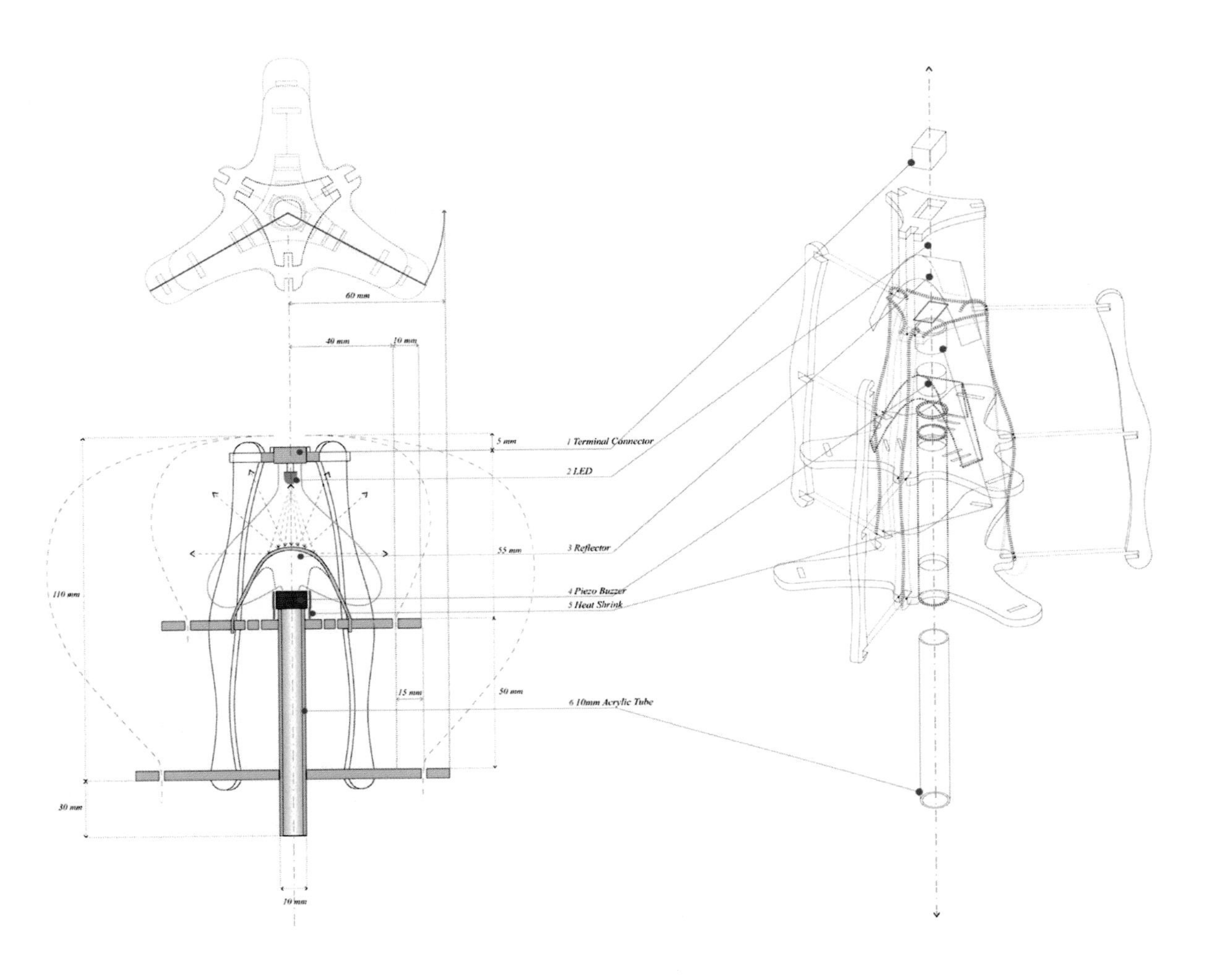

Figure 08.14
Responsive component diagram, METAfollies, Marco Poletto and Claudia Pasquero, ecoLogic Studio, FRAC Center in Orleans, France, 2013

range and performance capabilities of the piezo buzzers, their quantity is harnessed through looping time delays and augmented analogically by the length of acrylic tubes extending from the tendrils, producing infinite variations. The emergent complexity between materiality, responsive technologies, and interaction continues to build relationships between time, space, and materiality.

The *METAfolly* utilizes an "ambient" mode of response through establishing vague yet evolving relationships. By animating inanimate materials through human interaction, a relationship is drawn between our experience of "trashing" that questions how we value contemporary landscape. These relationships reference environmental phenomena and behavior through correlating human interaction with sounds that evoke the complexity of naturally occurring sounds found within the landscape through entirely artificial materials. The folly does not attempt to elucidate a particular phenomenon, rather the

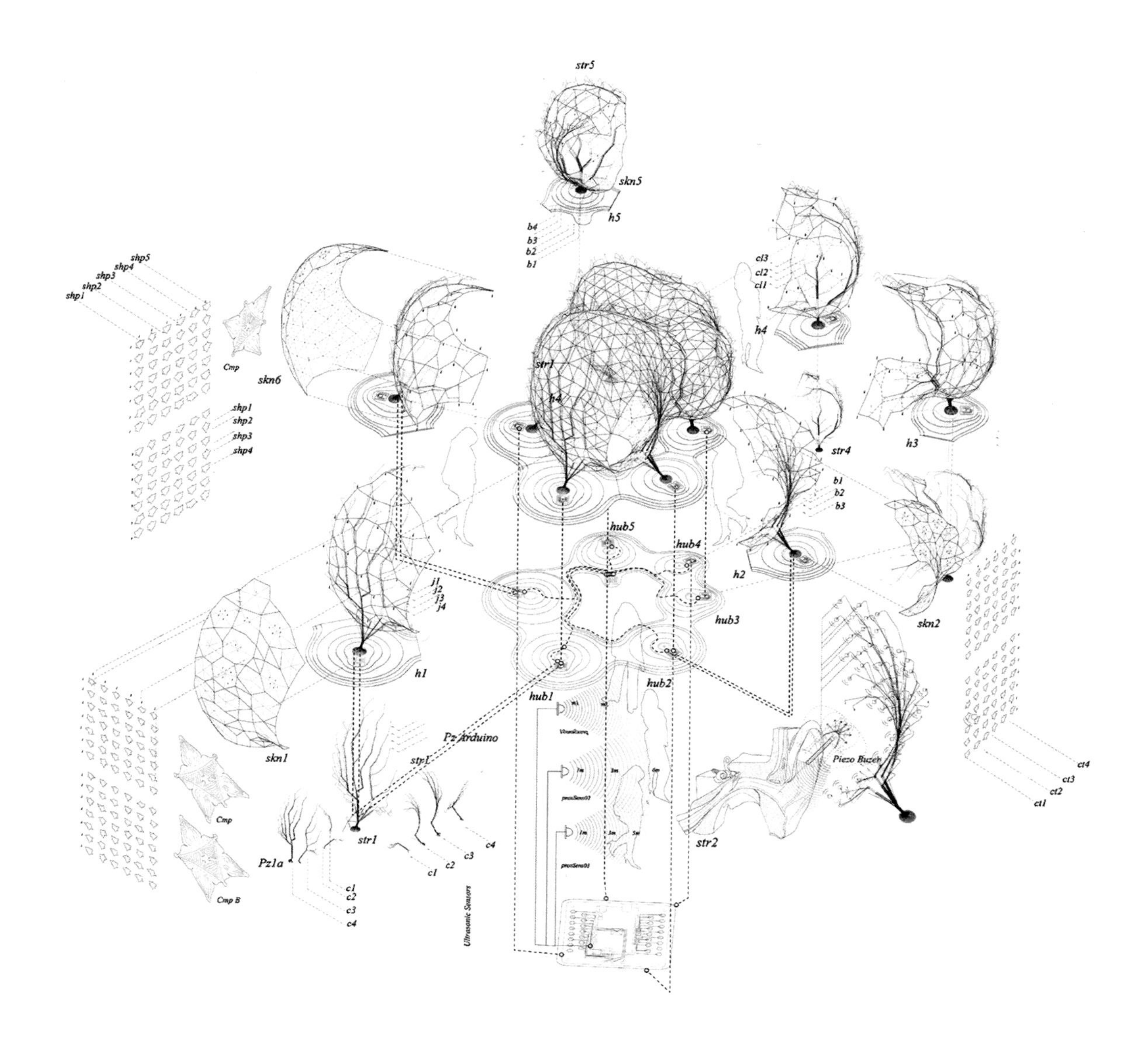

Figure 08.15 Assembly diagram, METAfollies, Marco Poletto and Claudia Pasquero, ecoLogic Studio, FRAC Center in Orleans, France, 2013

phenomenon (proximity) is abstracted, asking the viewer to develop an abstracted and metaphysical relationship with the installation—one that cannot be directly identified or choreographed. The translation of proximity into the orchestration of the piezo buzzers is not about establishing a specific connection, rather the swarming sounds produce a particular environment in which to experience the *METAfolly*. The translation of "the grotto" further relates the installation to an experience of landscape, and through this ambient interaction the installation takes on a new naturalness, heightening and exaggerating the experiential qualities of the *METAfolly*.

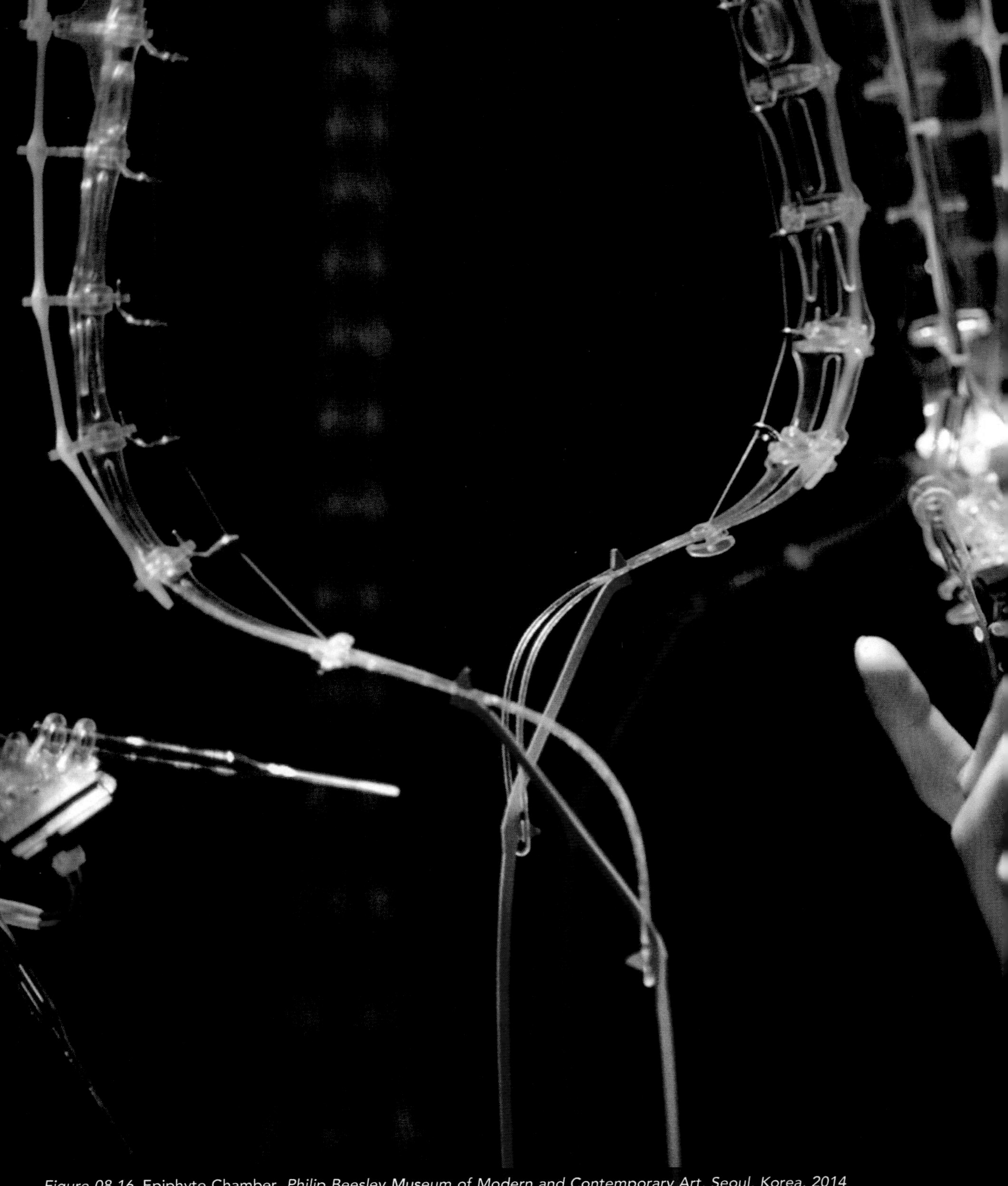

Figure 08.16 Epiphyte Chamber, *Philip Beesley Museum of Modern and Contemporary Art, Seoul, Korea, 2014*

Epiphyte Chamber

Seoul, Korea, 2014

Philip Beesley

Design team: Martin Correa, Andrea Ling, Jonathan Gotfryd

Collaborators: Rachel Armstrong, Brandon Dehart, Rob Gorbet. Contributors: Sue Balint, Matthew Chan, Vikrant Dasoar, Faisal Kubba, Salvador Miranda, Connor O'Grady, Anne Paxton, Eva Pianezzola, Sheida Shahi, May Wu, Mingyi Zhou

Scope: Temporary responsive installation at the Museum of Modern and Contemporary Art, executed with support from The Social Science and Humanities Research Council of Canada, NSERC CRSNG, Waterloo Architecture, Ontario Arts Council, and a continuation of the *Hylozoic Series*

Technology: Software: Arduino (Bare-Bones), Processing, C/C++, Rhino, Autocad, Adobe Suite

Hardware: Microcontroller boards, Atmel ATmegal68 microcontroller, custom daughter boards, sharp infrared proximity sensors, shape memory alloy, digitally fabricated components

---

The most recent of Philip Beesley's robust and highly developed series of responsive installations *Epiphyte Chamber* was selected for the inauguration of Korea's new National Museum of Modern and Contemporary Art complex in Seoul as a part of "The Aleph Project," an exhibition curated around the unique emergence of

Figure 08.17
Epiphyte Chamber, *Philip Beesley Museum of Modern and Contemporary Art, Seoul, Korea, 2014*

Figure 08.18
Construction of Epiphyte Chamber, *Philip Beesley*
Museum of Modern and Contemporary Art, Seoul, Korea, 2014

contemporary artworks in the twenty-first century responding to complex network theory. Developed through collaboration between theorists, designers, scientists, and artists, the works are meant to be performative and interactive. *Epiphyte Chamber* was selected to represent the future of architectural innovation, highlighting Philip Beesley's strategies for, "oscillating combinations of bottom-up and top-down design methods."[19] Beesley's work is regarded for using responsive technologies to emulate living systems. Through complex assemblies of thousands of lightweight fabricated components, the installations perform as metaphorical biological systems through exchanges with participants and surrounding environs.

The *Hylozoic Series* is conceptually modeled after "hylozoism," an ancient belief describing all matter as having life.[20] As the earliest form of anthropomorphism, hylozoism contended "that the various objects of the natural world are animated with life, intelligence, and passion similar to man's."[21] The installation takes on near-living qualities by performing actions expressed as "breathing," "caressing," and "swallowing" actuated in response to slight movements, airflow, and fluid exchanges.[22] The subtlety, complexity, and intricacy of this constructed environment masks the logic of the embedded and artificial intelligence, characterizing the mode of response as *ambient.* "Epiphyte" specifically refers to aerial plants that derive their nutrients only from the air. *Epiphyte Chamber* emulates these processes by continuing to explore a burgeoning sub-category within synthesized environmental architecture termed "chemical protocell metabolisms." The elaborate hanging web of delicate plastic fabricated pieces, microprocessors, and sacs of fluid delicately shifts over time, creating only "whispers" of movement. In Beesley's words, "The

Figure 08.19
Epiphyte Chamber, *Philip Beesley Museum of Modern and Contemporary Art, Seoul, Korea, 2014*

diffusive, dissipative form language described here offers a strategy for constructing fertile, near-living architecture."[23] The perceptual experience of this installation is not one of elucidation or direct connection, but an overall experience of qualities, descriptive of the system's functions emerging over time through subtle cues and correlations.

Philip Beesley challenges the permanence of responsive architecture with his long series of installations initially characterized by *Hylozoic Ground*. His designed structures operate as a system capable of responding to its surroundings but susceptible to an overall failure of the system. For Beesley, the idea of failure or "death" of the system instills "empathy" within the viewer.[24] Although still within the realm of perception, the concept of a truly impermanent system will result in architectures closer to "living" systems. The installation functions more as a bodily entity than as an ecosystem—transferring substances in, out, and within—without the ability to reorganize or evolve. To project into the future of this genre, the next step within the development of a synthetic system seeking to emulate living systems is for failure to become an adaptive capability—thus, moving from a synthetic system to a synthetic ecology.

NOTES

1 Eric Schlosser, *Fast Food Nation: the Dark Side of the All-American Meal*, (Boston: Houghton Mifflin: 2001).

2 Definition of "Ambient" Merriam Webster http://merriam-webster.com/dictionary/ambient.

3 Dunne, *Hertzian Tales*, 15 (see chap. 1, n. 18).

4 Peter Weibel, "Intelligent Ambience," Promotional leaflet for Ars Electronica, Linz: ORF, 1994. Cited by Dunne, *Hertzian Tales*, 15.

5 Charles Duhigg, *The Power of Habit: Why We Do What We Do in Life and Business* (New York: Random House, 2012).

6 Joost Van Loon, "Social Spatialization and Everyday Life," *Space and Culture* 5, no. 2 (2002): 88–95.

7 Zachary Pousman and John Stasko, "A taxonomy of ambient information systems: four patterns of design." *Proceedings of the Working Conference on Advanced Visual Interfaces*, New York: ACM, 2006. 67–74.

8 Will Kane, "Oakland cops aim to scrap gunfire—detecting ShotSpotter," SFGate (blog). March 14, 2014. Available online at http://sfgate.com/crime/article/Oakland-cops-aim-to-scrap-gunfire-detecting-5316060.php.

9 Mark Weiser and John Seely Brown, "The Coming Age of Calm Technology," 1996. Available online at http://ubiq.com/hypertext/weiser/acmfuture2endnote.htm. This is a revised version of Mark Weiser and John Seely Brown, "Designing Calm Technology," *PowerGrid Journal* 1, no. 1 (July 1996).

10 Andrew Galloway, "When Biology Inspires Architecture: An Interview with Doris Kim Sung," ArchDaily (blog), May 14, 2014. http://archdaily.com/?p=505016.

11 Doris Sung, "Bloom," dO|Su Studio Architecture (blog), http://dosu-arch.com/bloom.html.

12 Andrew Galloway, "When Biology Inspires Architecture."

13 See, Michael Fox and Miles Kemp, *Interactive Architecture* (chap. 1, no. 19) and Franziska Eidner and Nadin Heinich, *Sensing Space: future architecture by technology* (Berlin: Jovis, 2009).

14 Rob Ley and Joshua G. Stein, "Reef," Rob Ley (blog). http://rob-ley.com/Reef July 2015.

15 Philip Beesley and Omar Kahn, "Responsive Architecture/Performing Instruments," *Situated Technologies, Pamphlet 4: Responsive* (New York: The Architectural League of New York), 2009.

16 Claudia Pasquero and Marco Poletto, "METAfolly," ecoLogic Studio (blog), March 2, 2013. http://ecologicstudio.com/v2/project.php?idcat=3&idsubcat=66&idproj=120.

17 Ibid.

18 Ibid.

19 Philip Beesley, "Dissipative Models: Notes Toward a Design Method," in *Paradigms in Computing: Making, Machines, and Models for Design Agency in Architecture*, Eds. by David Jason Gerber and Mariana Ibañez, (Los Angeles, CA: eVolo Press, 2014), 33.

20 William A. Hammond, "Hylozoism: A Chapter in the Early History of Science," *The Philosophical Review* 4, no. 4 (July 1895): 394–406.

21 Ibid, 395.

22 Philip Beesley, "Introduction: Liminal Responsive Architecture," In *Hylozoic Ground: Liminal Responsive Architecture: Philip Beesley*, edited by Beesley, Philip, Hayley Isaacs, Pernilla Ohrstedt, and Rob Gorbet (Cambridge, MA: Riverside Architectural Press, 2010), 12–33.

23 Beesley, "Dissipative Models," 33.

24 Beesley and Kahn, "Responsive Architecture/Performing Instruments." In reference to Philip Beesley and Omar Kahn's discussion in which Beesley describes how the potential failure of the system instills empathy in the viewer/participant.

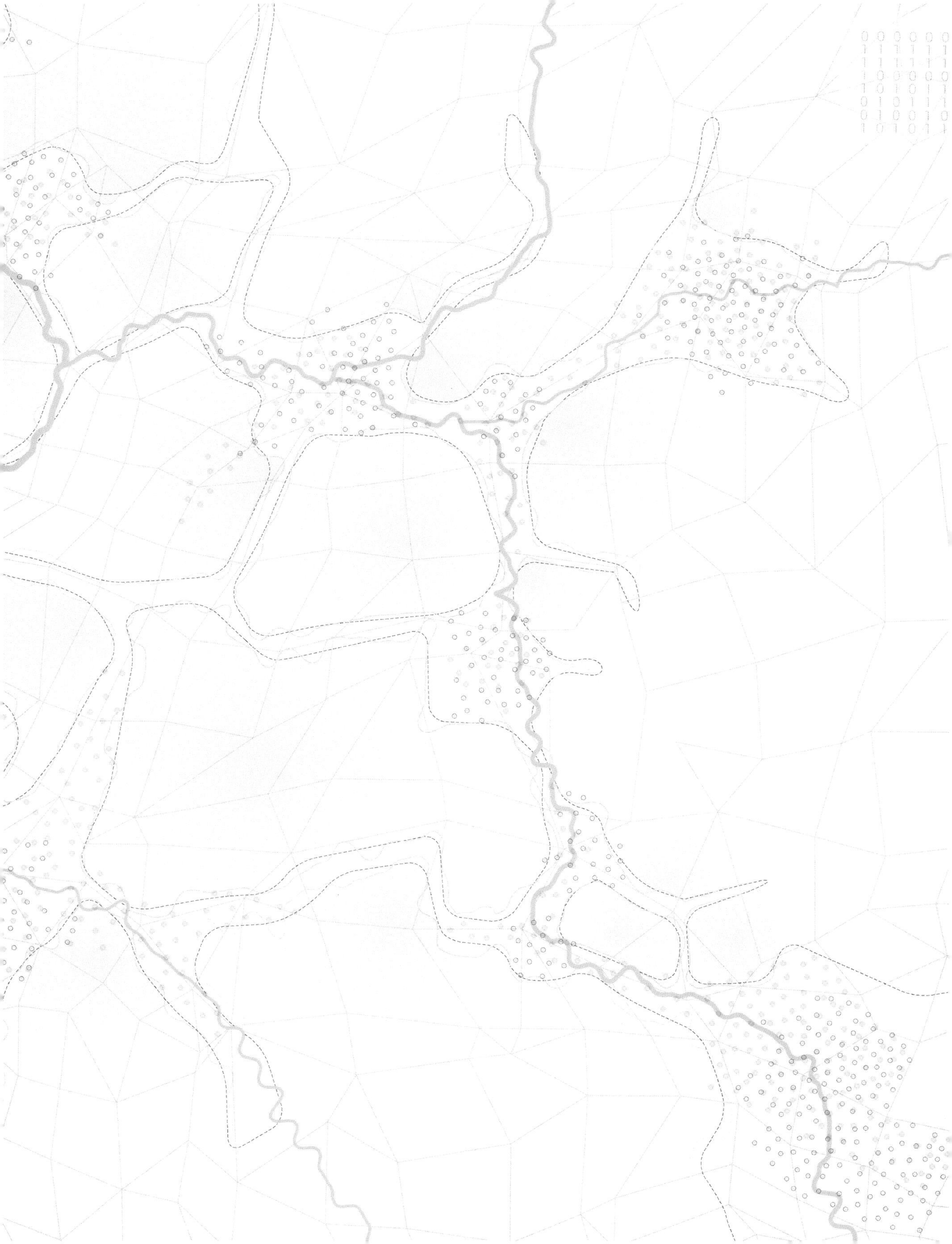

MODIFY

09

Modify reaches beyond virtualized or hybrid realities and presents the potential for responsive technologies to reshape environments. Modification of the environment is intrinsic to the discipline of Landscape Architecture and generally to human habitation. Intentionally or unintentionally, human actions are continually altering the environment across scales. *Modification* of the landscape, in the context of responsive landscapes, can be the product of a series of procedures that are linked to a specific logic. In the field of responsive architecture, these modifications are often the transformation of a facade morphology or the regulation of a thermal envelope. Within the landscape, these procedures are composed of hybridized logics to form interactions between biotic and abiotic entities. Examples of hybridized logics are seen at the planetary scale through the domestication of flora and fauna, at the continental scale through infrastructure and ecological maintenance, and at the territorial scale through contemporary agriculture. In the example of domestication, plant and animal species are selected over successive generations at the local scale through a logic that prescribes predictability, stability, and high yield. The landscape is transformed at varying speeds as these logics engage in a feedback loop, formatting the landscape to support these altered biologies.

Within the discipline of Landscape Architecture both the metaphor and application of computation present design methodologies to address hybridized landscapes, proposing designs as networks, operated by equally systemic protocols. This notion of computation facilitates indeterminate outcomes—which under the influence of repetition, recursion, and automated processing offer greater nuance with every iteration. Anastassia Makarieva considers the relationship of computation to systems ecology in her chapter "Cybernetics" in which she contrasts the inherent logics of biotic life to human computing capabilities: "biotic regulation can be viewed as an automatically-controlled operating system where the program of automatic control has been tested for reliability in an experiment lasting for several billion years."[1] Whereas, humans are wholly incapable of constructed technological and computational systems that amount to the logics operating across the landscape, "controlled by dozens of independently functioning unicellular and multicellular organisms, each living cell processing an information flux similar to that of a modern PC."[2] The combination of digital computation and biological computation further illustrates the potency of hybridized landscape logics.

Modifications are constantly and consistently occurring as a consequence to technological intervention and control. Over long periods of time, technological developments associated with the manipulation of the landscape are honed to become more reliable and precise. These logics exist in a multitude of different landscapes, from the urban to the infrastructural scale. A lawn in Los Angeles is the product of maintenance and an expression of societal norms. The lawn typology has a logic to how it is formatted and regulated. The aggregate impact of lawns across arid landscapes requires high amounts of water resources, fertilizers, and increased energy costs while the landscape of the lawn is maintained as an individual entity. This is a technological expression with specific logics that maintain a small green oasis in an otherwise arid climate. Similarly this logic could be expanded to imagine robustly managed systems that go beyond providing benefits based solely on human comfort and a cultural aesthetic.

The incredible infrastructural interventions in Coastal Louisiana have precipitated numerous control structures to mitigate further damages to the ecological system. The Caernarvon Freshwater Diversion structure, completed in 1991, diverts fresh water from the Mississippi River just below New Orleans to the east of the river. The diversion structure maintains a gradient of fresh to brackish to saline conditions in the Breton Sound estuary, delivering up to 8,000 cubic feet per second of freshwater to combat salt water intrusion. The diversion is monitored primarily by levels of salinity to establish a stable gradient or intermediate marsh across the basin, with special attention to oyster farming. The operation of the structure is informed by real-time monitoring within the basin, and then assessed against the water level on the river side of the levee and the basin side to regulate flow. The impacts of operational infrastructures, such as Caernarvon diversion, are authored by rule-based regulations supporting precise goals with measurable outcomes. As the fields of architecture and landscape architecture become more interested in the potential for flexible infrastructure to respond to ecological conditions. "What seems crucial is the degree of play designed into the system, slots left unoccupied, space left free for unanticipated development."[3] Stan Allen states that "Infrastructural systems work like artificial ecologies"—thus, their organizing principles and "degree of play" should be considered within the context of the broader landscape as it is synthesized through ecological metaphors.[4] Keller Easterling expands the potential role of infrastructure to include "the countless shared protocols that format everything from technical objects to management styles to the spaces of urbanism—defining the world as it is clasped and engaged in the space of everyday life."[5]

In perhaps a less obvious instance of landscape logics, Emma Marris uses the example of Yosemite to expose the management of the United States National Park system in her text *Rambunctious Garden*.[6] Yosemite, marketed as "wilderness," is maintained to replicate the moment it was preserved, a moment of transitioning American conceptions of nature linked to legislative protocols—dictating by law the management of different classifications of nature. The measures implemented to maintain stability can be understood as logics—from controlled water resources to the frenetic orchestration of bison populations. As notions of preservation, conservation, and general wilderness are transformed by contemporary conceptions of ecology, what we decide to manage and what we privilege as "wild" will undoubtedly yield new logics.[7] In his project for *Landscape Futures*, Liam Young speculates on the future of the Galapagos Islands, where "drone-animals" and "hybrid devices" become the curators of an environment that is better described as "an ornamental garden or historic park for ecotourists, an image of nature dramaturgically staged and populated by a cast of the augmented and genetically modified."[8] These three examples of landscape logics can be advanced to augment landscapes beyond their current bounds.

In *Marking Time: On the Anthropology of the Contemporary*, Paul Rabinow (influenced by John Dewey) explores notions of "the contemporary" as an assemblage of "fading" and "emergent" ontologies "maintained in motion and in time."[9] Although the text is targeted towards the discipline of Anthropology, it is quite applicable to developing an understanding or "inquiry" of the contemporary landscape or site—an assemblage of dynamic and temporal conditions existing through real-time perceptions that are too complex and interrelated to be rendered as an epoch. Similar to Charles Waldheim's provocation for the necessity of "distanced authorship" to facilitate an understanding of dynamic landscape processes over deterministic form,[10] Rabinow recognizes that understanding the current, "require[s] some critical distance from its own assumptions."[11] If the complexity of contemporary ecological systems can only be defined as a patchwork of ecological conditions shifting and switching into known and unknown landscapes, then current attempts to understand the contemporary ecological condition should take note of Rabinow's statement:

> . . . the question of how older and newer elements are given form and worked together, either well or poorly, becomes a significant site of inquiry. I call that site the contemporary.[12]

Responsive Landscapes has offered five methods in the preceding chapters for utilizing responsive technologies as a form of "inquiry" to frame or ask questions about the contemporary landscape without "deducing it beforehand."[13] The contemporary landscape, as a site of emergent and novel conditions, undoubtedly holds conditions that have escaped previous attempts of translations. Landscape phenomena, captured by sensors and instruments, offer opportunities to understand the contemporary through responsive modes of observation, impossible to be structured or conformed to conventional modes of analysis in which a particular question exists at the outset. A nuanced understanding of the interactions within ecological systems may require this method of inquiry in order to address emergent landscape qualities beyond their novelty.[14]

Similarly influenced by John Dewey, and his instrumental theory of knowledge,[15] Brian Davis has been developing a theory of "landscapes as instruments" as they address the "gap between intention and reality."[16] In considering landscapes similar to those exhibiting hybridized logics, Davis makes a compelling argument for the instrumentality of matter that reaches beyond Bennett's concept of "vital materialism," to incorporate objects, knowledge, and ideas as operative instruments, in which "the designer is actively engaged with a host of other objects in the activities of knowing and acting in a specific situation."[17] While *modify* (and this text in general) is focused on the application of responsive technologies (a notably obvious form of instrumentalism), it is contingent on the operative material qualities of the landscape as well as the feedback between their intentional and unintentional logics.

Stewart Pickett, a leading scientist and proponent for humans as components of ecosystems, studies urban ecological systems with the intent to bridge urban and ecological design. In their paper, "Urban Ecological Systems," Pickett et al. outline ecological approaches for exploiting the design of urban ecologies. While traditional ecological models may prioritize inputs and outputs, rendering the system as a "black box" within urban ecological systems entities should be considered for their specific agency:

> . . . the structural details and richness of processes that take place within the boundaries of the system are a major concern of contemporary ecosystem analysis. Contemporary ecosystem ecology exposes the roles of specific species and interactions with communities, flows between patches, and the basis of contemporary processes in historical contingencies.[18]

While the argument for ecological, non-deterministic strategies seemingly holds the answers to the problems of the contemporary landscape, it then becomes problematic to find methods that actually resist deterministic outcomes. At what point are goals, scaffolds, and protocols actually open-ended? The intent of *modify* is not to translate an existing condition or to construct a condition—rather, through direct and indirect behavioral adjustments, it suggests recursive inquiry through repetitive action. With this nuanced environmental sensitivity, slight or automated adjustments become the mediators of human intentionality with the environment.

A critical moment in this discourse is represented by Chris Reed's paper, "The Agency of Ecology," where he outlines strategies for designing indeterminate and ecological landscape systems. In his description of "*analog ecologies*" he proposes that "the responsive behaviors of living systems in nonliving constructions or processes" have the potential to influence constructed ecologies "to react to changing inputs and to adapt their nature to the new or revised condition at hand." Reed even uses Chuck Hoberman's *Adaptive Fritting* project—a responsive architectural skin and early case study for responsive architecture to describe the potential to apply this methodology—of responsiveness to landscape:

> we might think of the design of flexible social spaces: physical scaffolds for the playing out of open-ended (but not limited) social and cultural—as opposed to ecological—activities. And in large-scale, complex urban projects, we might imagine the setting up of responsive administration frameworks; "if, then" scenarios; and management strategies that allow for feedback loops, input, and responsiveness over time.[19]

These modes of inquiry and modification must be met with an intense criticism—recognizing the inherent polemic of embedding design solutions as solely methodology—heightened by the idolization of indeterminism. The current discourse for "conceptualizing and operationalizing the city"[20] are robust with strategies that ambiguously oscillate between moments of intention, distance, and autonomy.

Shannon Mattern describes the computational and operative city as, ". . . one that can be hacked and made more efficient—or just, or sustainable, or livable—with a tweek to its algorithms or an expansion of its dataset." She articulates this polemic well, as she points out the shared "faith in instrumental rationality" or "solutionism"[21] bolstered by "a tendency toward data fetishism and methodolatry, the aestheticization and idolization of method" in which the optimism for logics lacks the counterweight of the tangible.[22] While *Responsive Landscapes* proposes attempts resembling "methodolatry," the foundation of this work recognizes that this mode of operationalizing the landscape is already happening. The introduction of computational and operational mechanisms into the landscape are exaggerating existing hybridized logics, which have been shaping and modifying the landscape for centuries. If landscape architecture intends to engage the issues of the contemporary landscape, more aligned with the infrastructure of globalization, there must be a fierce engagement with the mechanisms of current operational paradigms.

Speculation from proclaimed futurists is a critical generator for the development of these strategies. In the same way that breaking apart robotic toys for prototyping responsive objects is critical for proof of concept and the beginnings of dialogue with technological advances, this extreme sci-fi speculation—in which it may only exist through drawings, film or artistic expression—formulates future realities that take inaccessible technologies into wholly new realities by challenging their application as provocation.

> What landscape futures might we explore when entire terrains are internally mechanized, given partial sentience, and able to interact with one another wirelessly, over great distances of time and space?[23]

The rapid speeds of technological development alongside landscape manipulation must be met with an equally rapid study of how landscape architecture as a discipline can begin to engage the types of "wicked problems" we seek to solve.[24] The application of responsive technologies to articulate hybrid realities needs to be reconsidered and harnessed for purposes outside the long histories of technological development for control of dynamic systems. These hybrid realities create a series of questions that are important for the profession as we confront embedded forms of computation in landscape systems. How can we deconstruct the relationships of current operative technologies with ecosystems? What forms of sensing and monitoring are necessary to readily engage landscape systems at local and global scales? How does this pervasive form of computation move from control to more active methods of engagement?

The projects in *Modify* present a series of methodologies that hint at methods of tuning, nudging, resisting, and choreographing, which counter historical techno-centric paradigms of control. While the projects in *Modify* can be seen as technophilic, at their core they represent a nascent speculation of intelligence, resistance, and micro adjustments in which the tuning and nudging of landscapes requires interventions to co-evolve as they learn about the contexts in which they reside. This form of modification does not learn to find the correct solution, but instead learns to stay relevant to an evolving context. Slight modifications become the language of slowly constructing and deconstructing landscapes to push the environment into new trajectories. In the same light, modes of resistance have the potential to act in situations of disturbance or failure, drawing from a body of learned interactions to shift into new states. Choreography shapes outcomes through a sequencing of bifurcations as it simultaneously tests new expressions. These applications for modification are central to the current discourse in landscape architecture and are a prompt for designers. The field of *Responsive Landscapes* and the application of responsive technologies will be a critical tool for addressing the most pressing issues within the profession.

Open Columns
Buffalo, New York, 2007

Liminal Projects

Collaborators: Omar Khan and Laura Garofalo

Scope: Temporary responsive installation

Technology: Carbon Dioxide Sensors and Custom Fabrication

Open Columns is an installation comprised of flexible columns that can change their surface morphology based on an environmental input. The columns modify space as they physically contract into or extend down from the ceiling. The columns are constructed from a perforated and pliable rubber that allows them to physically transform. Because of the particular dimension and range of extension, the unfolding and contracting of the columns alters the physical accessibility of the space. Although the columns are non-supporting, they take on faux structural characteristics, perceptually altering the building structure but primarily modifying the spatial layout.

The project confronts issues of time that are important to interaction designers; specifically, the relationship between an action and the change that is registered. The choreography of a response to a specific environmental phenomenon can be choreographed in many ways to create perception of association, disconnection, or synchronicity. In *Open Columns*, the choreography is non-synchronous in the short term, creating a response that seems to come from a logic that is internal to the column. The change in the column happens at a much slower speed than the occurrence of the phenomena, thus creating a delay. The installation uses carbon dioxide as the trigger for the morphological change of the column: when a change in carbon dioxide is detected, the columns begin their expansion or contraction. This effect allows for an action to begin–for example, the gathering of people who are emitting carbon dioxide—and then begins to alter that behavior through the slow expansion of the column.

The alteration of time states makes for an interesting relationship between sensing the environment and the response within the device. The modification in the environment is not a one-to-one relationship but instead creates a life within the object: letting it respond to the input and asking the environment to then respond back to the object. This formulates a symbiosis between occupants of the space, the devices themselves, and the biological sources of carbon dioxide. As an input, carbon dioxide could come from any source—making the installation respond in a multitude of ways, and require multiple

Figure 09.02 Columns contracted within exhibition space. Open Columns, Omar Khan and Laura Garofalo, Liminal

experiences with the behavior before tying it back to environmental phenomena. In this sense the architecture itself serves as a diagram of the relationship between material, atmosphere, and users: a changing vector that alters itself and remixes spatial conditions.[25]

Open Columns also confronts the threatening nature of responsive or reactive architectures—the concept that the structural envelope is in motion. As a prototype, the columns use rubber as a material that is perceptually unthreatening and has properties of transformation. The columns are physically *soft* and the speed of the transformation is perceived as minimal, slowly unfolding without abrupt starts and stops.

Open Columns, while primarily an architectural investigation, creates a compelling case study for how morphological changes within space may not have one-to-one temporal relationships to the sensed phenomena. In many ways this is how landscapes function: they evolve based on environmental inputs such as a leaf opening and repositioning itself perpendicular to the sun throughout the course of the day. These responses are directly tied to environmental change

Figure 09.03 Columns expanded, Open Columns, *Omar Khan and Laura Garofalo, Liminal Projects, Buffalo, NY, 2007*

Figure 09.04
Column suspension system, Open Columns, Omar Khan and Laura Garofalo, Liminal Projects, Buffalo, NY, 2007

but temporally decoupled—ultimately creating learned relationships. As people are exposed to the behaviour, they may associate the expansion and contraction to a multitude of events: people gathering, an increase in the volume of sound, or possibly an increase in motion—all attributes that would likely increase as more carbon dioxide is emitted. As a method of modification this form of dissociated response and modification is an important way to understand environmental interactions.

Smart Highway

Netherlands, 2013

Studio Roosegaarde and Heijmans Infrastructure

Collaborators: Daan Roosegaarde

Scope: Prototype development for a collection of innovative responsive designs for regional scale highway signage, safety, and energy conservation

Technology: Photo-luminizing paint treatment, Motion detection, Induction charging, and small wind turbines

---

Smart Highway is a current design strategy envisioning intelligent and responsive methods for transforming the future of highway signage. Shifting the focus of innovation from the car to the highway, designer Daan Roosegaarde in collaboration with builder and developer Heijmans, re-imagines highway infrastructure as responsive and intelligent. The *Smart Highway* is designed to be responsive to light, weather conditions, and interaction to produce a sustainable, safe, and intuitive driving experience. A range of designs was developed to be embedded into existing highway infrastructure, replacing static and antiquated solutions—from *Glow-in-the-Dark* and *Dynamic Paint* for road surface markings responsive to light and temperature to an Induction Priority Lane capable of charging an electric car in motion. Glowing Lines have been implemented in the Dutch Province of Noord-Brabant while other aspects of the network are prototyped.

Composed of multiple responsive designs operating as a network—the *Glow-in-the-Dark Road, Dynamic Paint, Interactive Light, Induction Priority Lane*, and *Wind Light—Smart Highway* envisions a network capable of modifying its operational logic in response to changing specific traffic situations. Information elucidated through the network not only provides a readable and articulate landscape, but it also prompts the driver to make "smarter" decisions, modifying typical driving habits in response to real-time conditions.

Glow-in-the-Dark Road applies a special photo luminizing powder as a treatment to painted road markings, lighting up the markings at night for up to a ten hour period, reducing the need for added lighting and reflectors. Similarly, *Dynamic Paint* responds to seasonal weather conditions related to dangerous driving conditions. The paint is envisioned to glow in iconic snowflake patterns across the road when it is cold, icy, and slippery, elucidating precisely where the dangerous road conditions are in real-time to eliminate risk. The driver is then able to modify the operation of their vehicle in response to this information.

Figure 09.05 Rendering of Glow-in-the-Dark Road, Smart Highway, *Studio Roosegaarde*

continuous line
dotted line

Figure 09.06 Rendering of Wind Light, Smart Highway, *Studio Roosegaarde*

Figure 09.07
Rendering of Interactive Light, Smart Highway, *Studio Roosegaarde*

Figure 09.08
Rendering of Induction Priority Lane, Smart Highway, *Studio Roosegaarde*

Figure 09.09
Rendering of Dynamic Paint, Smart Highway, *Studio Roosegaarde*

Interactive Light, *Induction Priority Lane*, and *Wind Light* all function to reduce the amount of energy required to light highways by conserving light for necessary conditions. *Interactive Light* addresses light pollution and energy waste by recognizing that light is really only required along the road when cars are present. The road edge is lined with directed lights projecting across the road surface that respond in multiple levels of brightness using motion sensors to illuminate in proximity to moving cars. Wind Light populates the edges of the highway with small wind turbines equipped with individual LEDs whimsically brightening in response to windy conditions and passing cars.

The *Induction Priority Lane* and *Dynamic Lines* are perhaps the most futuristic visions of the network. Given the trending—and necessity of—transitioning to electric cars, the *Induction Priority Lane* both privileges and supports electric car owners while providing a sustainable source of energy to the car in transit. It is unclear how Daan Roosegaarde and Heijmans intend to materialize these performative designs.

Smart Highway was conceived for the Dutch landscape; however, it has international potential to reshape highway infrastructure by adding, removing, and crafting individual design innovations to reflect the needs of a specific landscape by considering climate, weather, transportation, traffic, density, and energy needs. A *New York Times* article about *Smart Highway* connects the work of Roosegaarde and Heijmans to research on intelligent roadways across the world. In the United States, *Solar Roadway* is working "streets made of electricity-generating photovoltaic cells that generate enough electricity to recharge electric cars at rest and parking lots." In Madrid, Via Inteligente is embedding wireless technology into paving stones to "integrate Wi-Fi and Bluetooth connectivity in durable paving tiles" to facilitate autonomous vehicles. And in Germany, experiments along the autobahn include overhead electric cables for "long-distance, zero-carbon transport" at the infrastructural scale.[26] By addressing the operational needs of the driver in relation to phenomena affecting specific driving conditions in situ, *Smart Highway*'s network functions with higher levels of efficiency and safety through intelligent decision making.

Vacuolar Effluvia Genesis

Atchafalaya Basin, Louisiana, 2011

Josh Brooks and Kim Nguyen, LSU Responsive Systems Studio, Fall 2011

Scope: Design of Ecological Devices Operating within the Atchafalaya Basin, Louisiana

Technology: Arduino, Grasshopper, Physical Prototyping

Vacuolar Effluvia Genesis (VEG) is a speculative project exploring the technologies that facilitate the re-sequencing of ecological processes within the Atchafalaya Basin, Louisiana. *VEG* was designed in the Responsive Systems Studio, an advanced topic studio taught by Bradley Cantrell (author) and Frank Melendez. The project specifically targets algae with a containment system that is designed to autonomously sequester algal blooms during decomposition. The proposal addresses the overabundance of nitrate fertilizers that are entering the Atchafalaya Basin from anthropogenic sources in the upper Mississippi River watershed. In co-ordination with efforts up river to minimize inputs to the watershed, *VEG* focuses on the eutrophication cycle within the basin. As an infrastructure, *VEG* modifies the ecosystem as it modifies specific moments of the algal life cycle and tunes itself with moments of high nutrient discharge.

Eutrophication is the oscillation in aerobic microbial decomposition and amounts of dissolved oxygen within the water column. As pollutants are introduced to the basin, particularly nitrate fertilizers, there is an increased growth of suspended algae. This increased growth creates issues during the decomposition of the algae as it descends into the water column and is consumed by microbes. The microbes also consume oxygen, leading to a higher rate of decomposition and lower amounts of dissolved oxygen producing a condition known as hypoxia. Hypoxic conditions are the leading cause of fish kills, producing areas known as "dead zones" where oxygen-consuming aquatic life cannot survive.

The project attempts to redefine the asymptote between dissolved oxygen and microbial composition to maintain a more balanced ecosystem. The team developed a method of sequestration that seeks to make localized interventions throughout the basin and monitoring the effect in real-time. This process allows for micro-changes throughout the process and the amalgamation of these units creates a scaled response to fertilizer inputs. Sequestration occurs when the autonomous units detect 5ppm of dissolved oxygen and actuate in order to alter the processes associated with the overabundance of nitrates.

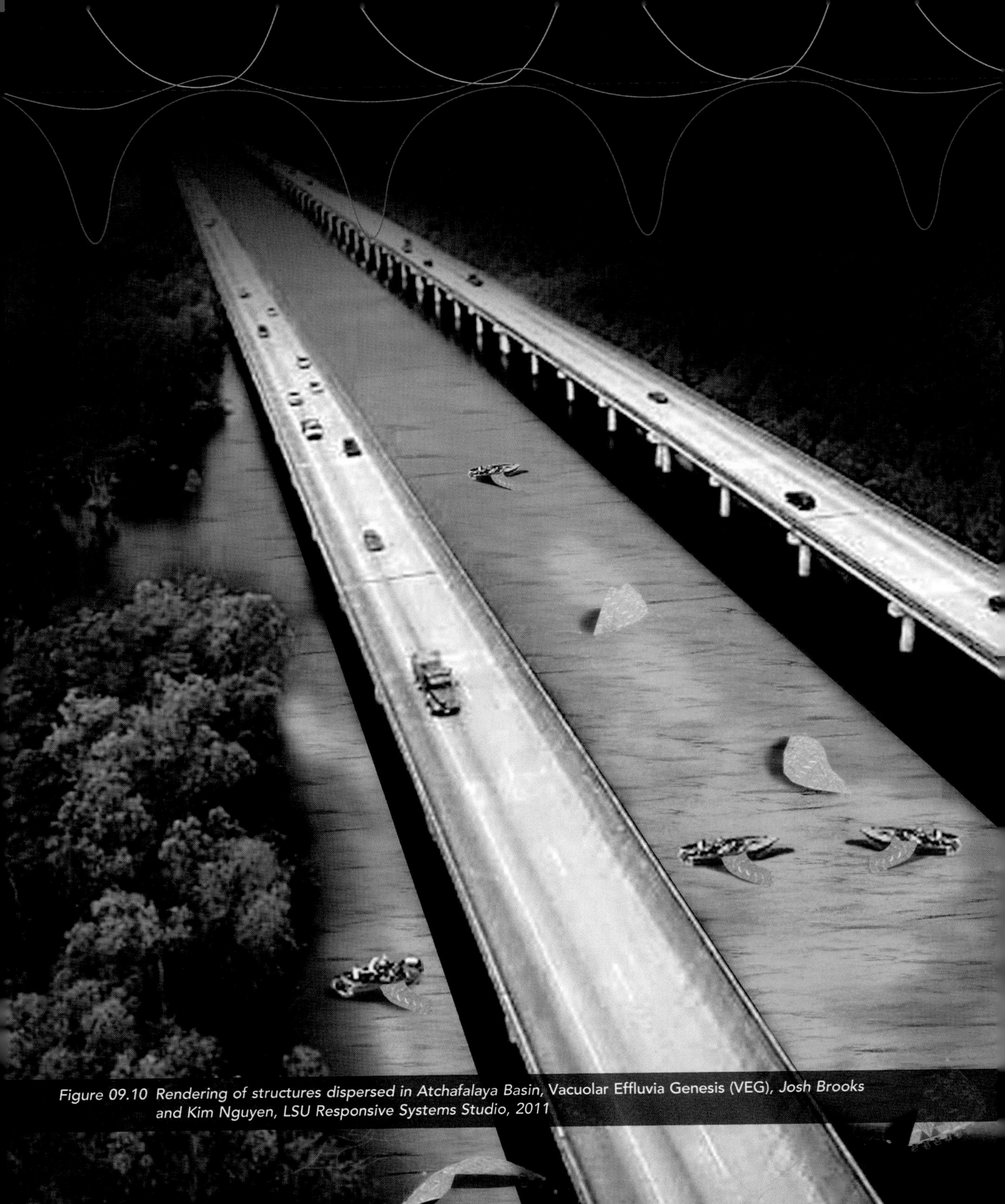

Figure 09.10 Rendering of structures dispersed in Atchafalaya Basin, Vacuolar Effluvia Genesis (VEG), Josh Brooks and Kim Nguyen, LSU Responsive Systems Studio, 2011

Figure 09.11 Rendering of tubing system and inflatable skin, vacuolar Effluvia Genesis (VEG), Josh Brooks and Kim Nguyen, LSU Responsive Systems Studio, 2011

Figure 09.12 Rendering of inner components of inflatable skin and relationship to water surface, Vacuolar Effluvia Genesis (VEG), Josh Brooks and Kim Nguyen, LSU Responsive Systems Studio, 2011

Figure 09.13 Close up of inflatable skin, Vacuolar Effluvia Genesis (VEG), *Josh Brooks and Kim Nguyen, LSU Responsive Systems Studio, 2011*

The devices use a system of tubes, resembling roots, dispersed throughout the water column to collect algae saturated water and hold the solution during the decomposition process. After a typical decomposition process, lasting approximately 14 days, the solution produces three by-products: biogas, mineral matter, and water. The gas is collected into an inflatable dome structure composed of a series of pockets, and the mineral matter and water are deposited back into the water column. After repeated cycles the dome fills, creating a harvestable biogas used to fuel the device or for external needs. The domes produce approximately 500 liters of gas per 1,000 liters of algae-saturated water and the fuel can be converted to butanol, used to power gasoline engines.

VEG proposes a method of responsiveness that redefines infrastructure as a distributed and tuneable network of devices. The modifications that are made to the environment are incremental and take into account not only local responses but large-scale issues upstream, in the interior of the continent. This form of modification to the environment poses interesting questions about the instrumentality of local responsive devices to alter large-scale ecosystem health. How can these devices be tuned locally and balance local and territorial data? What are the logics that are employed to create emergent infrastructures?

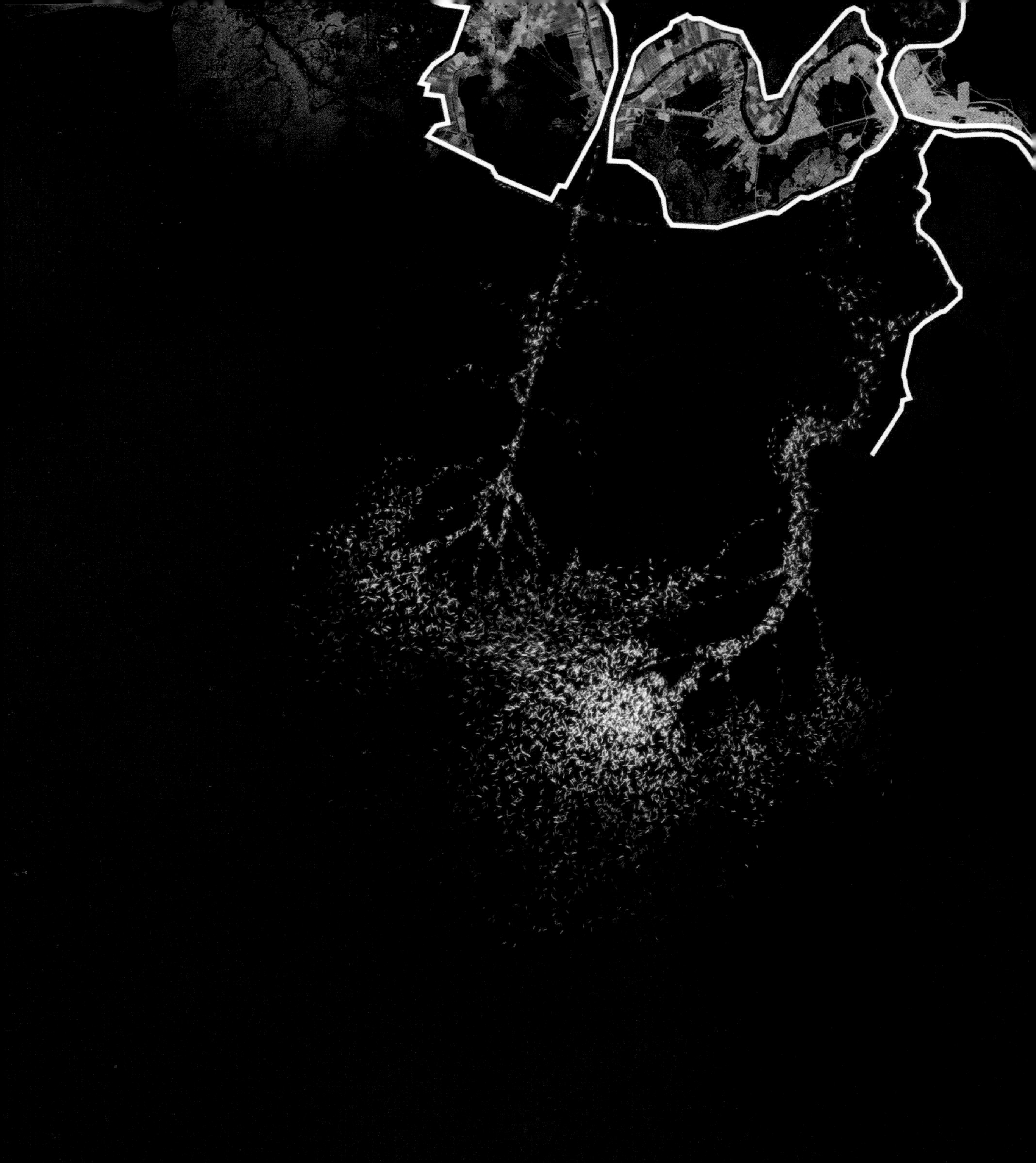

Figure 09.14 Pods dispersed at the mouth of the Atchafalaya and Wax Lake Deltas, Pod Mod, *Charlie Pruitt, Brennan Dedon, and Robert Herkes, Responsive Systems Studio, 2011*

Pod Mod

Atchafalaya Basin, Louisiana, 2011

Charlie Pruitt, Brennan Dedon, and Robert Herkes, LSU Responsive Systems Studio, 2011

Scope: Design of Ecological Devices Operating within the Atchafalaya Basin, Louisiana

Technology: Arduino, Grasshopper, Physical Prototyping

The reordering or manipulation of single attributes often drastically alters the potential of larger systems. The specific gravity of sediment in relation to the velocity of water determines if sediment will remain suspended in the water column or will settle—water moving quickly carries heavier sediment. The *Pod Mod* proposal exploits buoyancy as a method of extending the sediment carrying capacity of water through the alteration of existing infrastructure. *Pod Mod* is a speculative design proposal addressing issues of sedimentation, dredging, and land building at the mouth of the Atchafalaya Basin, developed in the Responsive Systems Studio, taught by Bradley

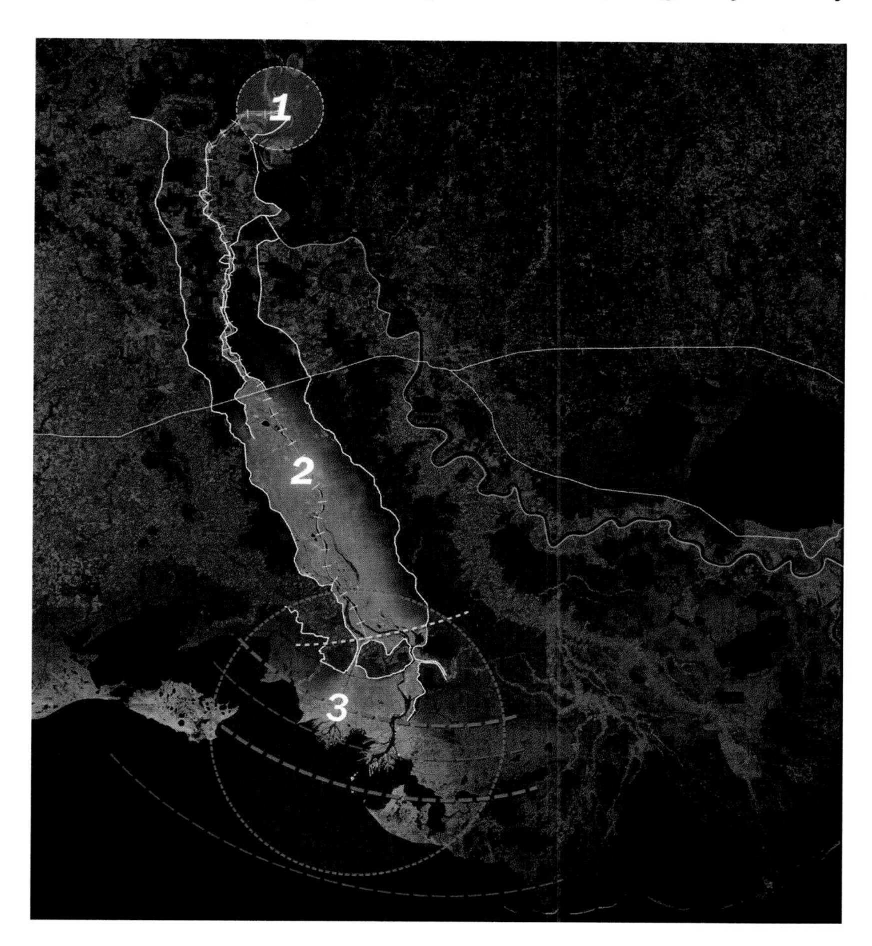

Figure 09.15
Areas of slow velocity in the Atchafalaya Basin, Pod Mod, *Charlie Pruitt, Brennan Dedon, and Robert Herkes, Responsive Systems Studio, 2011*

Figure 09.16
Area of intervention at the Low Sill Structure, Pod Mod, Charlie Pruitt, Brennan Dedon, and Robert Herkes, Responsive Systems Studio, 2011

Cantrell (author) and Frank Melendez. The project proposes the creation of sediment "pods" equipped with an inflatable bladder and embedded with responsive materials, designed to float from the top of the Atchafalaya Basin towards the Gulf of Mexico. By enabling the sediment to overcome low velocities within the basin and nudging the sediment to settle in optimum areas, land is built at the mouth of the Atchafalaya River and Wax Lake Delta—two of the only places that land is currently building within Coastal Louisiana—faster and more robustly.

The Atchafalaya Basin contains the Atchafalaya River, a major distributary of the Mississippi River and the shortest route to the Gulf of Mexico. The Atchafalaya River, as the shortest route to the Gulf of Mexico, is the preferred path for the Mississippi River. In defiance, the Old River Control Structure finished in 1963, keeps the Mississippi flowing towards the city of Baton Rouge and New Orleans. In times of high water, the basin serves as the overflow for the Mississippi River and is the largest swamp in the United States—one of the most ecologically rich areas in the world. The Old River Control Structure,

Figure 09.17
Pods floating down the river, Pod Mod, Charlie Pruitt, Brennan Dedon, and Robert Herkes, Responsive Systems Studio, 2011

built at the junction of the Atchafalaya and Mississippi Rivers, diverts 30 percent of the Mississippi River's water flow and 65 percent of its sediment flow into the Atchafalaya River. In the Gulf of Mexico at the lower end of the basin, the Atchafalaya River discharges into the Atchafalaya Bay via the Wax Lake Delta and the Atchafalaya Delta. Because there is an excess flow of sediment and nutrients, the Wax Lake and Atchafalaya Deltas are creating new land formations while the rest of the Louisiana coast is in decline. Even with this land building the majority of the sediment that is diverted into the Atchafalaya Basin is lost to deposition within the basin, never reaching the coast, or is propelled beyond the coast to settle in the depths of the Gulf of Mexico.

The project proposes to engage this process through the introduction of a sediment transport system at the Old River Control Structure. The system would consolidate the existing sediment load and convey it downriver through launches of sediment pods that would autonomously navigate the Atchafalaya River and deposit near the isohaline at the Gulf of Mexico. In addition to providing a concentrated sediment load, releasing the packaged sediment at the fluctuating isohaline actuates a robust series of landforms to develop at the edge of the current delta system. The sediment transport system is composed of two elements: an extrusion module that would be fitted within the existing Old River Control low sill structure and the pod, comprised of a mesh that holds the sediment and a biodegradable ballonet. The extrusion module would take advantage of the existing sediment load and direct the sediment into the biodegradable pods using the flow of the river water to push the sediment. The second element, the pod, is inflated to provide buoyancy. The pods

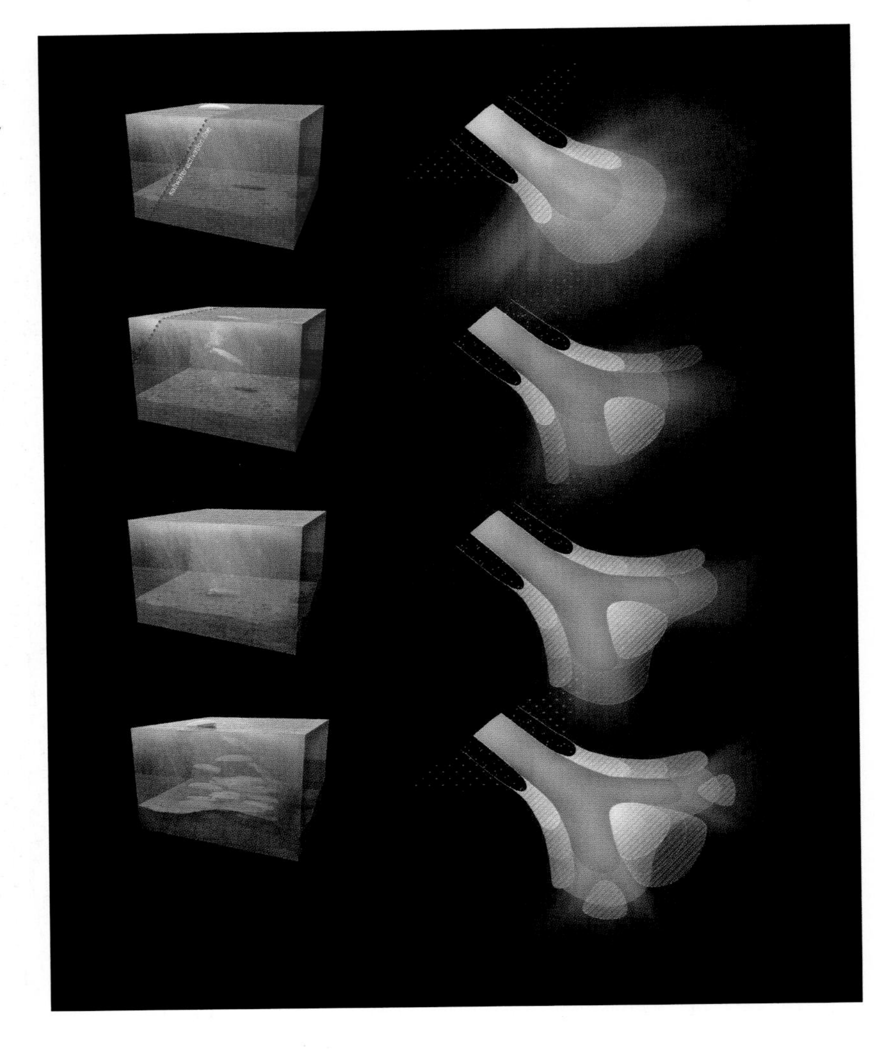

Figure 09.18 Isohaline causing Pods to deflate and sink, Pod Mod, Charlie Pruitt, Brennan Dedon, and Robert Herkes, Responsive Systems Studio, 2011

are sealed using a metal fastener that is highly susceptible to galvanic corrosion, designed to fail when exposed to the brackish saltwater at the isohaline. The failure of the fastener causes the pods to deflate and fall to the delta floor, creating a stabilized landform held together by the biodegrading pod. The material response of the fastener, tied to the dynamic phenomena of the mixture of salt and freshwater, creates land building at the leading edge of the delta. Each pod would also contain a small RFID tag that would give each deposition a unique identifier. As the pods are released they would be recorded, and by scanning the land formations from passing boats or through autonomous drones they would be re-identified in their resting location. The tracking of the pods would develop a point cloud of locations that is tied back to the launches of each volley of pods and can be used to better understand river dynamics, launching patterns, and success rates.

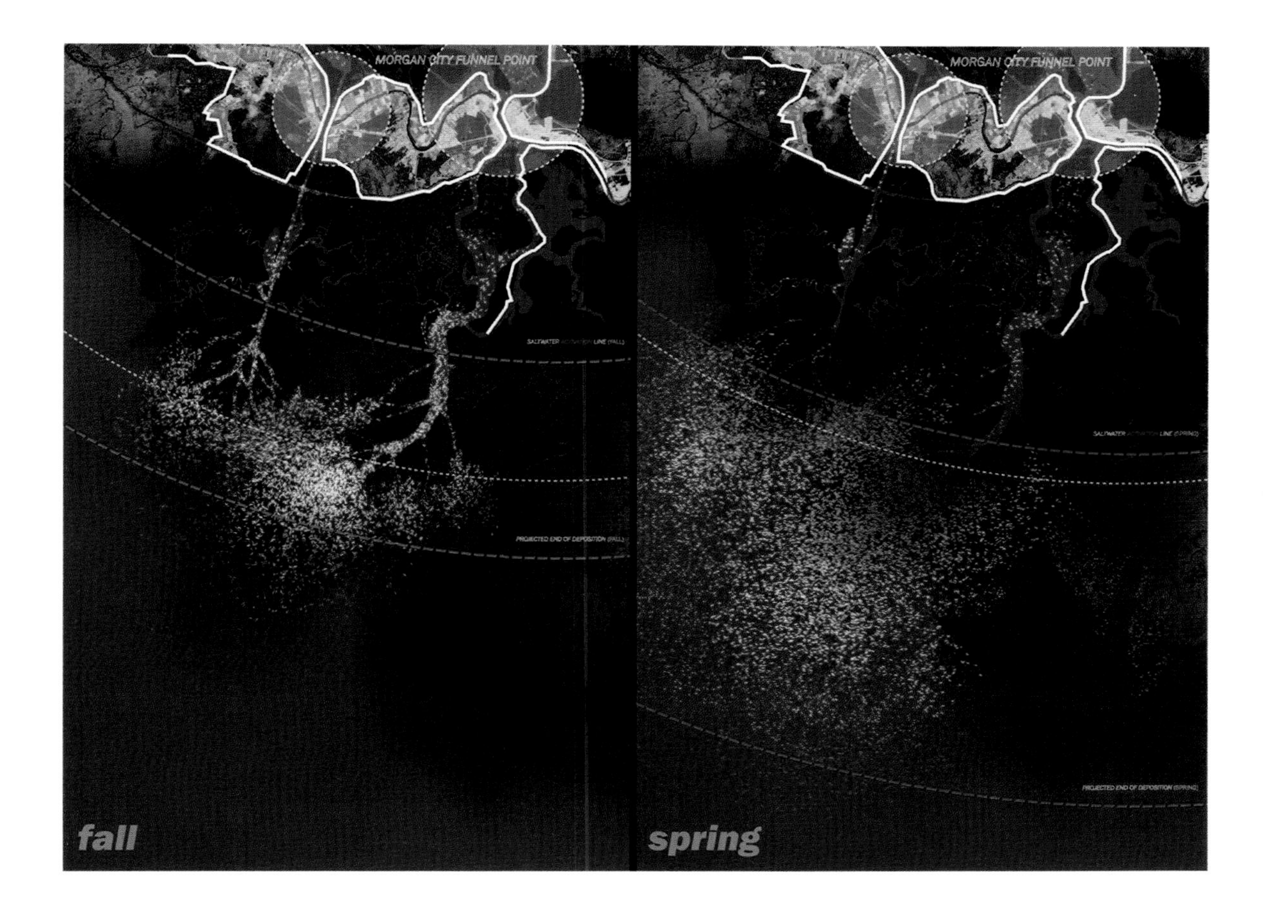

Over time, the process of deposition will build land south of Morgan City at the Wax Lake and Atchafalaya Deltas at an expedited rate, maximizing ecological habitat and storm surge protection. Because the sediment will travel farther down the Atchafalaya River, it will reduce the amount of dredging needed to maintain navigation within the basin. As a modifier of the landscape, the proposal picks up on an extremely subtle but important form of ecological choreography to form a highly responsive delivery system that is integrated within existing infrastructure. The system is modified by a composite of inputs, including navigational priorities, species interaction, or seasonal storm cycles. The pods become identifiers of sediment transport and promoters for overcoming issues of low velocity within the basin.

Figure 09.19
Dispersion zone based on fluctuating isohaline, Pod Mod, *Charlie Pruitt, Brennan Dedon, and Robert Herkes, Responsive Systems Studio, 2011*

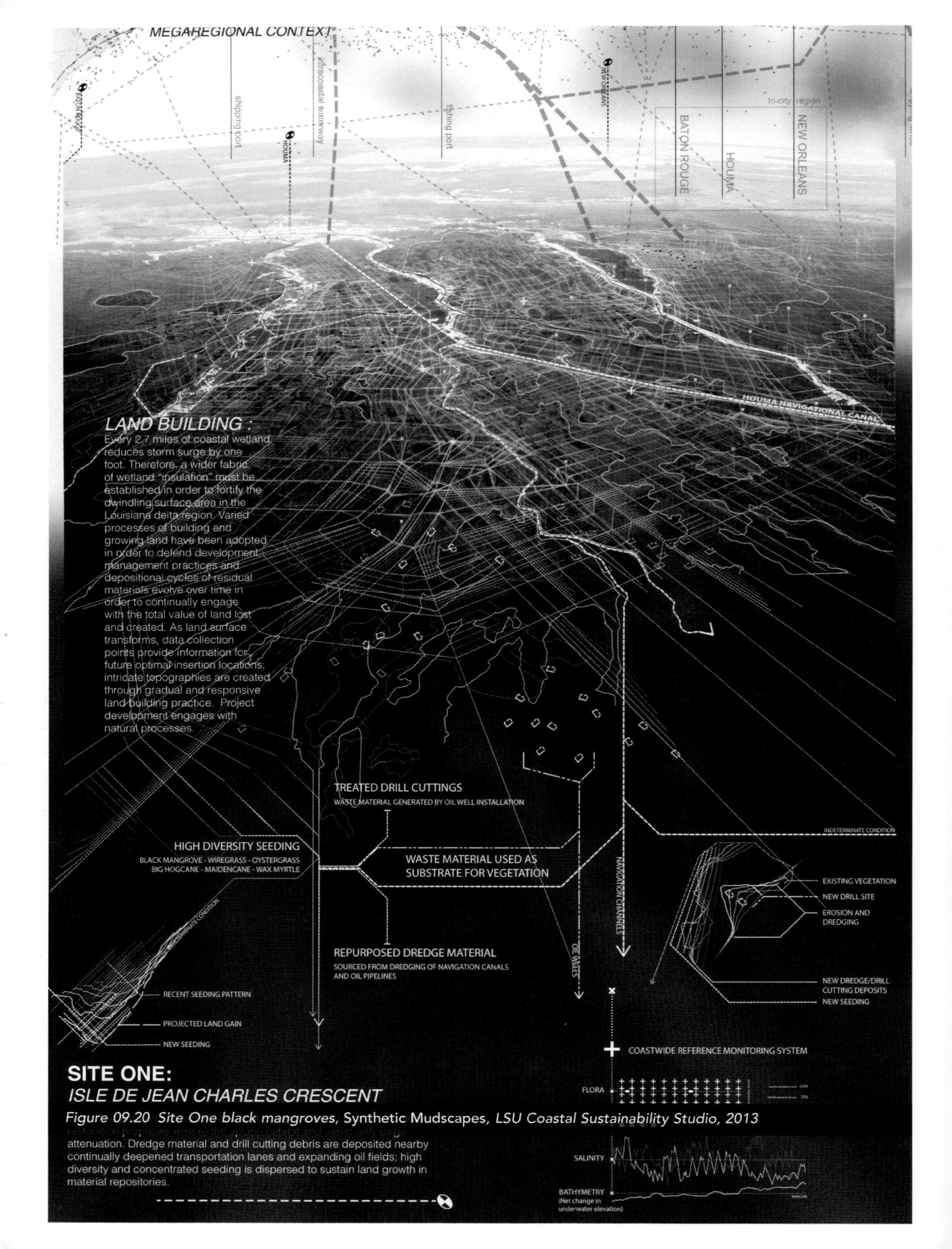

Figure 09.20 Site One black mangroves, Synthetic Mudscapes, LSU Coastal Sustainability Studio, 2013

Louisiana State University Coastal Sustainability Studio

Collaborators: Jeff Carney, Bradley Cantrell, Liz Williams, Matthew Seibert

Scope: Mississippi River Delta

Technology: Design Speculation on the Implementation of Real-time Modeling of Coastal Conditions

Synthetic Mudscapes

Mississippi River Delta, Louisiana, 2013

Synthetic Mudscapes is a speculative project at the scale of the Mississippi River Delta, imagining the reintroduction of deltaic fluctuation and strategic deposition of fertile material as an adjustable system, finely tuned through the expansive intensification and reinvention of an existing coast wide monitoring system within a responsive network. This project envisions this system as a virtual mesh allowing the landscape to be increasingly known and managed.[27]

In an attempt to defend the infrastructure, economy, and settlement of the gulf coast region, *Synthetic Mudscapes* proposes the fortification of a rapidly disappearing landscape by constructing new land to mitigate increasing risk. Land building is one in a range of viable strategies for long term storm protection and ecological fitness within the Louisiana delta. The *Synthetic Mudscapes* project envisions a

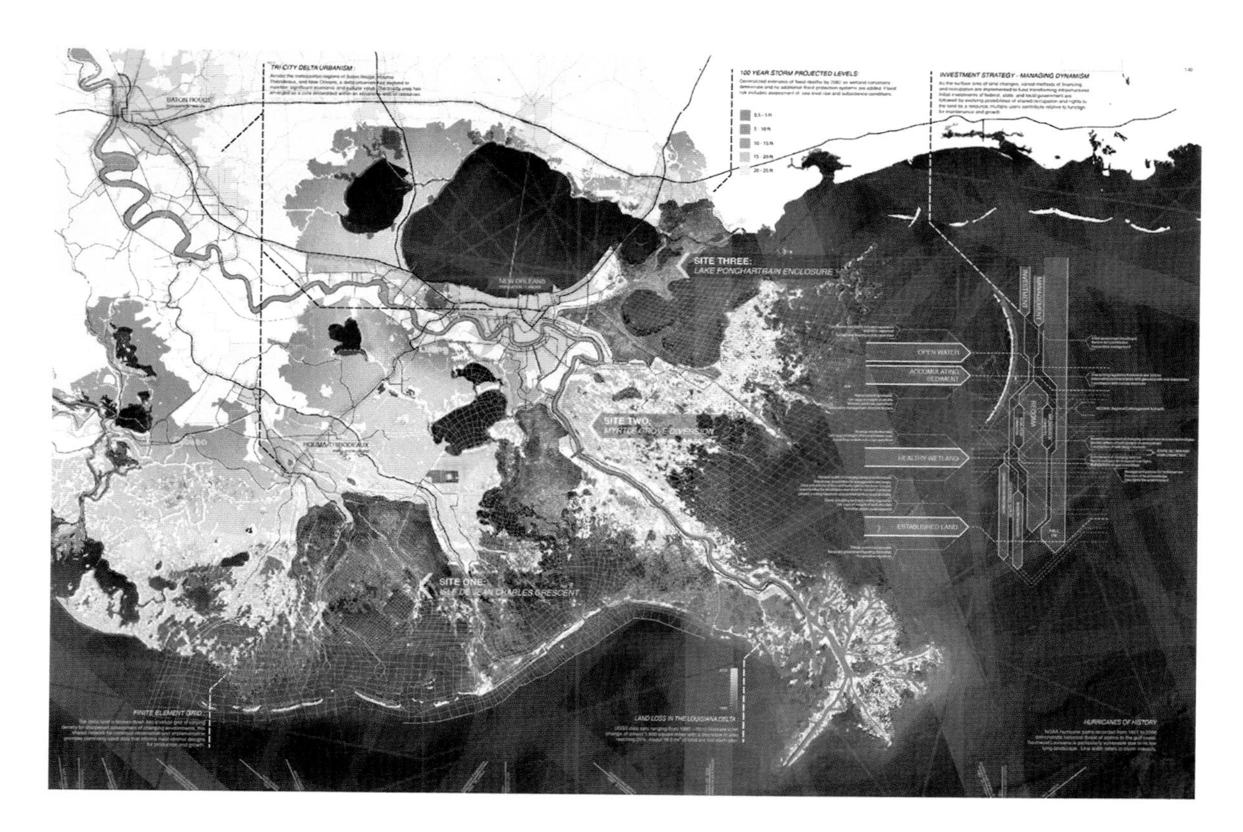

Figure 09.21
Synthetic Mudscapes, *LSU Coastal Sustainability Studio, 2013*

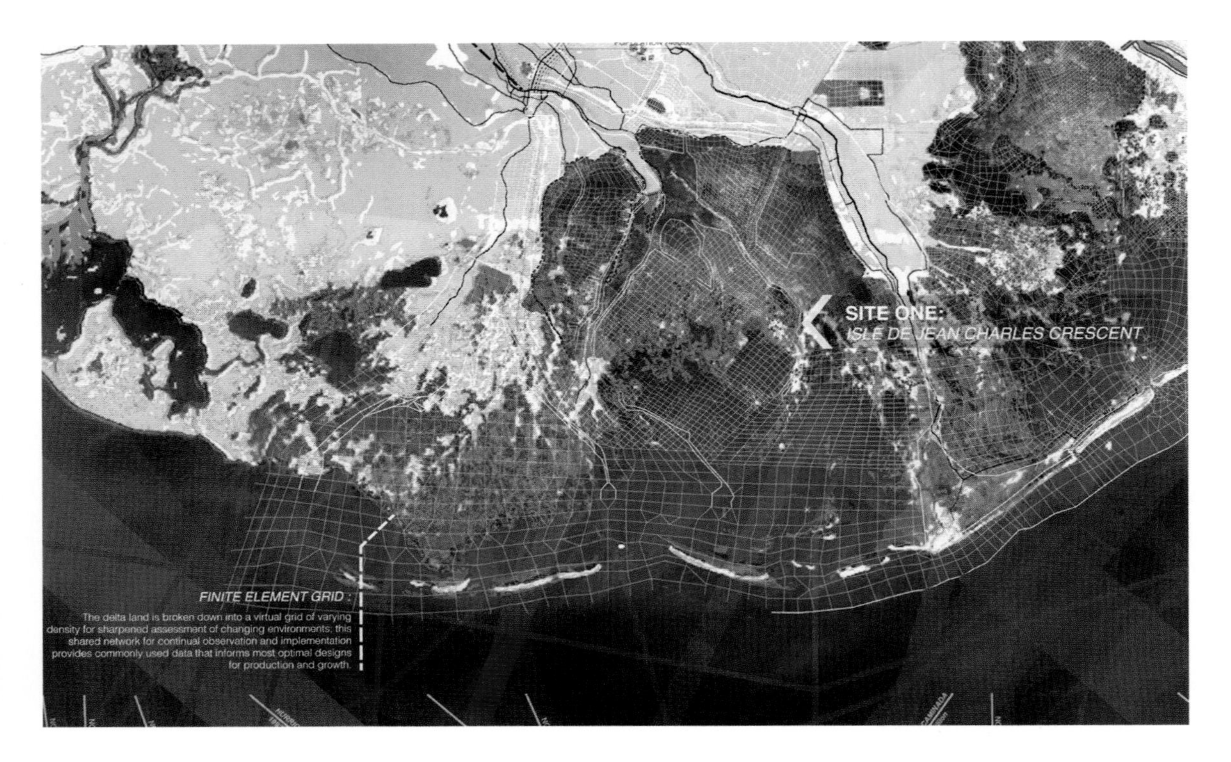

Figure 09.22
Finite element grid, Synthetic Mudscapes, *LSU Coastal Sustainability Studio, 2013*

future where the current incongruity of vast wetlands and a competing network of human settlement is transformed to reflect the inter-dependence needed for the future health of both domains. Focusing solely on the links between urban environments and economic drivers have constrained the dynamic delta environment for generations, and subsequently undermined the ecological fitness of the entire region. Change and indeterminacy are denied by traditional inhabitation through the implementation of monolithic infrastructure and static management regimes. The project proposes an alternative, where human occupation effectively strengthens environmental resilience in the face of sea level rise, land subsidence, and increased storm frequency.

The resulting synthetic landscape—part *wild*, part engineered—challenges conventional notions of urban and nature, rendering a future where the margin or "hinterland" has clear, measurable value, "to give place to such out-of-control, adaptive, robust, self-directed designs is to allow, or to install, a degree of wildness within them."[28] The reintroduction of deltaic fluctuation and strategic deposition of fertile material to form the foundations of a multi-layered defense strategy undermines the marginality of the coastal edge. The reframing of this marginalized condition through the lens of management casts a new system of values on the region. The use of simulation as a method of understanding historical data, making decisions in

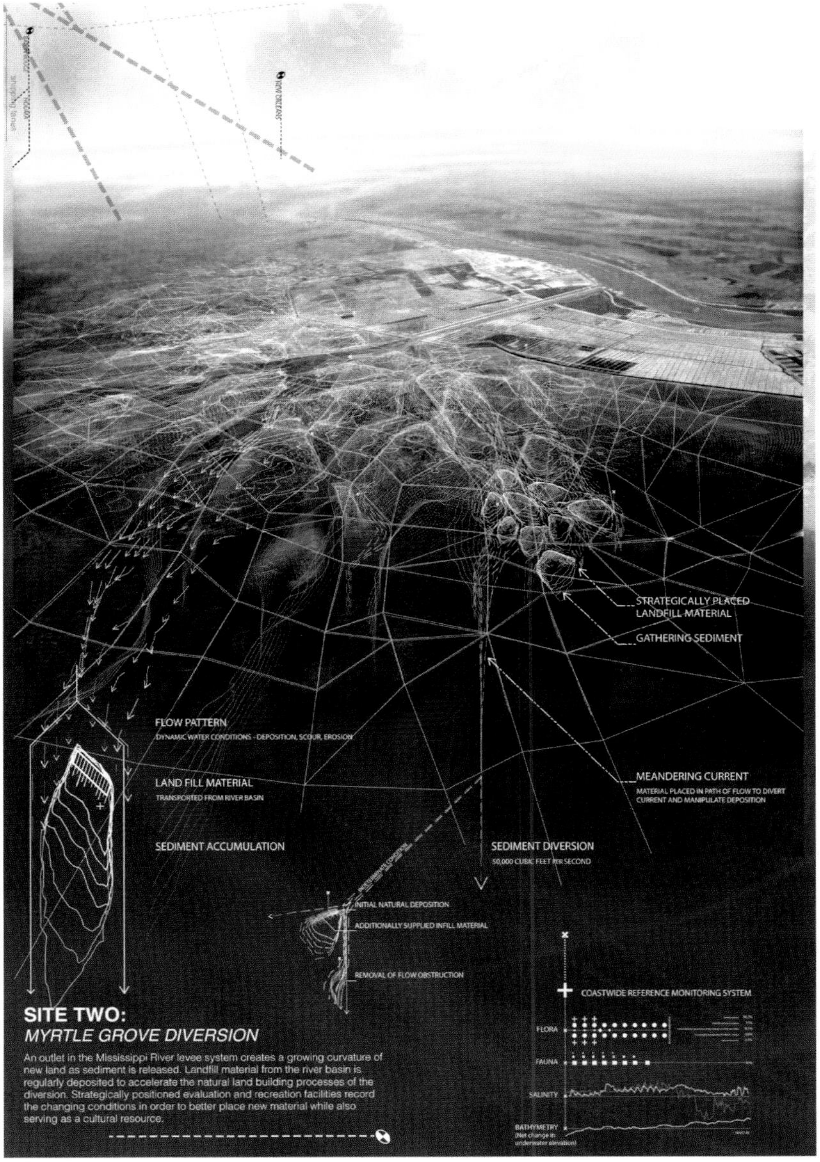

Figure 09.23
Site Two Diversion, Synthetic Mudscapes, *LSU Coastal Sustainability Studio, 2013*

the present, and projecting for long term decision making virtualizes the delta and forms a perceived infrastructure operating in tandem with an evolving ecosystem. This method is an evolved exploration of Sanford Kwinter and Cynthia C. Davidson's Chapter "Wildness" in which they determine, "[d]esign today must find ways to approximate these ecological forces and structures, to tap, approximate, borrow, and transform morphogenetic processes from all aspects of wild nature, to invent artificial means of creating artificial environments."[29]

Through the expansive intensification and reinvention of an existing coast wide monitoring system, a real-time sensing network gathers data to render a virtual mesh creating a landscape that is increasingly *known* and managed. This virtual mesh, a simulation of fluid and ecological processes, is the framework for recursive decision making and management within the newly introduced coastal land building infrastructure. The mesh is the current fidelity of current computational modeling of the deltaic systems at any given time and it can be assumed it will increase in fidelity in the future. What is not assumed in the project is that these simulations and models are an inherent truth; instead, they are seen as a series of layers that are considered in processes of decision making and response. At the scale of inhabitation, *Synthetic Mudscapes* attempts to choreograph regional ecosystem strategies of deposition and integration. The management methodologies aim to promote an assimilation of data that challenges the rigidity of urban margins and creates an interwoven framework of cultural and regulatory relationships.

As a strategy the project aims to embed many of the themes discussed within *Responsive Landscapes*. Most importantly, *Synthetic Mudscapes* is a proposal that does more than visualize complex phenomena and decidedly engages a new paradigm of ecological inhabitation. The direct connections between the products of human inhabitation are incorporated and processed through synthetic forms of land building, waste treatment, and resource extraction. The ecologies that are formed are novel landscapes that aspire to a high level of ecological fitness and resilience through an oscillating regime of maintenance and evolution. It is this type of response that implicitly forms a relationship between the device and the ecosystem, forming a computational symbiosis that not only manages a system, but begins to evolve with the ever changing needs of the system at the local scale and the effects at a larger scale. What are the logics that are employed to create emergent infrastructures? It proposes a form of emergent resistance that creates a symbiosis based on performance metrics that are never maximized for any of the participants.

Figure 09.24 (facing page) Site Three Recycle, Synthetic Mudscapes, *LSU Coastal Sustainability Studio, 2013*

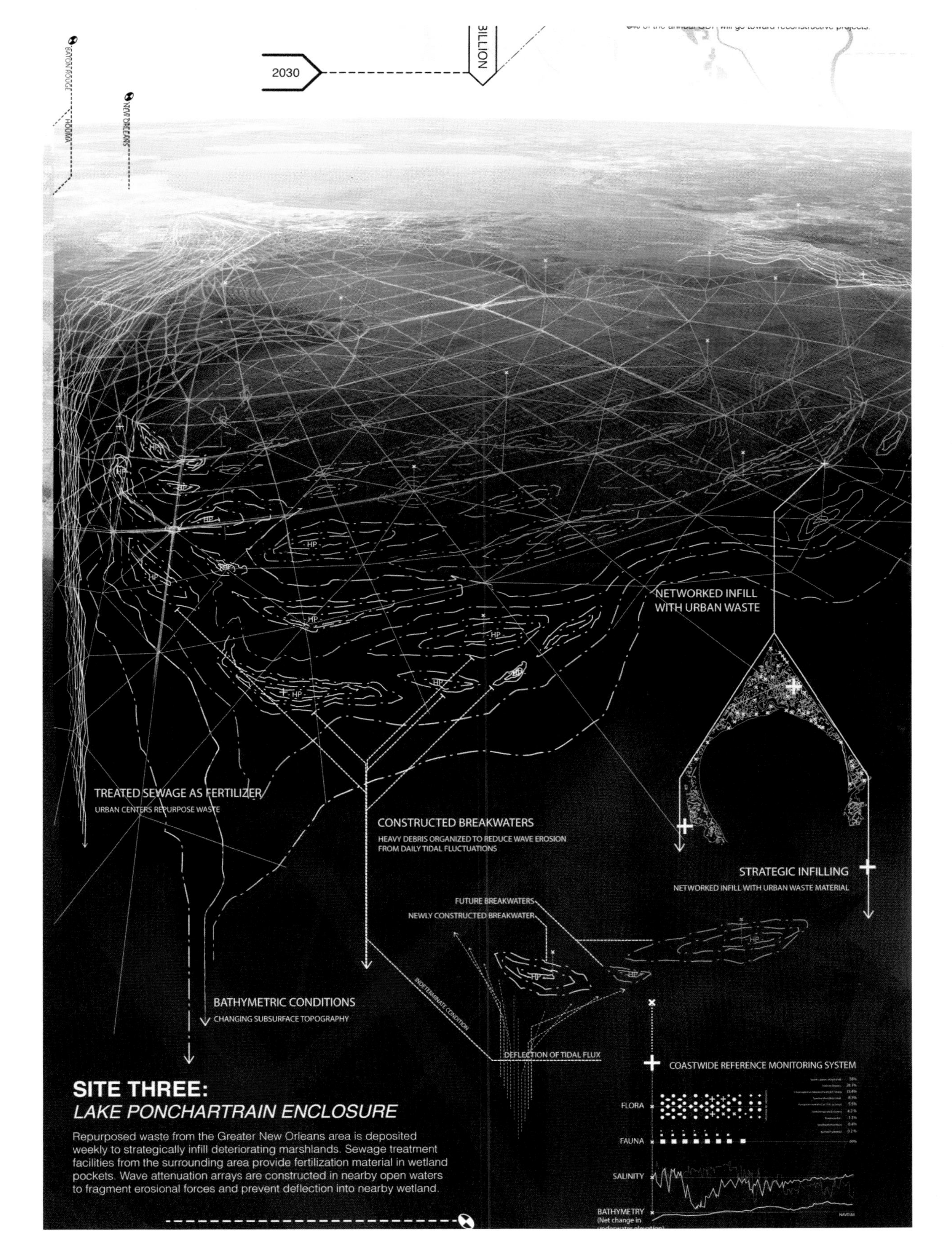

2030
BILLION
BATON ROUGE
HOUMA
NEW ORLEANS
NETWORKED INFILL WITH URBAN WASTE
TREATED SEWAGE AS FERTILIZER
URBAN CENTERS REPURPOSE WASTE
CONSTRUCTED BREAKWATERS
HEAVY DEBRIS ORGANIZED TO REDUCE WAVE EROSION FROM DAILY TIDAL FLUCTUATIONS
STRATEGIC INFILLING
NETWORKED INFILL WITH URBAN WASTE MATERIAL
FUTURE BREAKWATERS
NEWLY CONSTRUCTED BREAKWATER
INDETERMINATE CONDITION
BATHYMETRIC CONDITIONS
CHANGING SUBSURFACE TOPOGRAPHY
DEFLECTION OF TIDAL FLUX
COASTWIDE REFERENCE MONITORING SYSTEM
FLORA
FAUNA
SALINITY
BATHYMETRY (Net change in
SITE THREE:
LAKE PONCHARTRAIN ENCLOSURE
Repurposed waste from the Greater New Orleans area is deposited weekly to strategically infill deteriorating marshlands. Sewage treatment facilities from the surrounding area provide fertilization material in wetland pockets. Wave attenuation arrays are constructed in nearby open waters to fragment erosional forces and prevent deflection into nearby wetland.

NOTES

1 Anastassia Makarieva, "Cybernetics," in *Systems Ecology*, Vol. 1 of *Encyclopedia of Ecology*, Eds. Sven Erik Jørgensen and Brian D. Fath (Oxford: Elsevier, 2008): 806–812.

2 Ibid.

3 Stan Allen, "Infrastructural Urbanism," *Cambridge Journal of Architecture: Scroope* 9 (1997): 55.

4 Ibid., 57.

5 Keller Easterling, "Fresh Field," in *Coupling: Strategies for Infrastructural Opportunism*, Ed. Neeraj Bhatia (New York: Princeton Architectural Press, 2011), 10.

6 Emma Marris, *Rambunctious Garden: Saving Nature in a Post-wild World* (New York, NY: Bloomsbury, 2011), 17.

7 Rob Holmes, "New American Wilderness," Paper presented at 2014 CELA Annual Conference, Baltimore, Maryland (2014, March 27).

8 Manaugh, *Landscape Futures*, 47 (see chap. 4, n. 10).

9 Paul Rabinow, *Marking Time: On the Anthropology of the Contemporary* (Princeton, NJ: Princeton University Press, 2008), XI.

10 Charles Waldheim, "Strategies of Indeterminacy in Recent Landscape Practice," *Public* 33 (Spring 2006): 86.

11 Rabinow, *Marking Time*, 2.

12 Ibid., 2–3.

13 Ibid., 3.

14 Rabinow in reference to John Dewey, *Logic. The Theory of Inquiry*, Vol. 12, *The Later Works, 1925-38*, (Carbondale: Southern Illinois Press, 1991).

15 Specifically, John Dewey, *The Quest for Certainty: A Study of the Relation of Knowledge and Action* (New York: Minton, Balch and Company, 1929).

16 Brian Davis, "Landscapes and Instruments", *Landscape Journal* 32, no. 2 (2003): 161–176.

17 Ibid., 167.

18 Stewart Pickett, T.A., et al., "Urban Ecological Systems: Linking Terrestrial Ecological, Physical and Socioeconomic Components of Metropolitan Areas," *Annual Review Ecological Systems* 32 (2001): 127–157.

19 Chris Reed, "The Agency of Ecology," in *Ecological Urbanism*, Eds. Mohsen Mostafavi and Gareth Doherty (Baden Sweden: Lars Müller, 2010), 327–328.

20 Shannon Mattern, "Methodolatry and the Art of Measure," *Places Journal* (November 2013). Accessed 13 Feb 2015. https://placesjournal.org/article/methodolatry-and-the-art-of-measure/

21 Ibid., Mattern's critique of "solutionism" is related to an interview with Evgeny by Natasha Dow Schüll, "The Folly of Technological Solutionism: An Interview with Evgeny Morozov" Public Books (blog), September 9, 2003. http://publicbooks.org/interviews/the-folly-of-technological-solutionism-an-interview-with-evgeny-morozov.

22 Ibid.

23 Manaugh, *Landscape Futures*, 44.

24 Gale Fulton, "Landscape Intelligence," *Journal of Landscape Architecture* 31 (April–June, 2011): 46–53.

25 Fuller and Haque, "Urban Versioning System 1.0," 47 (see chap. 2, n. 13).

26 Paul Hockenos, "Street Smarts: From Holland, Bright Ideas for Highways," *New York Times*. April 28, 2013, New York edition. http://nytimes.com/2013/04/28/automobiles/from-holland-bright-ideas-for-highways.html?_r=1

27 Bradley Cantrell, "Synthetic Mudscapes," in *Paradigms in Computing: Making, Machines, and Models for Design Agency in Architecture*, Eds. David Jason Gerber and Mariana Ibañez (Los Angeles, CA: eVolo Press, 2014), 357–361.

28 Sanford Kwinter and Cynthia C. Davidson, *Far from Equilibrium: Essays on Technology and Design Culture* (Barcelona: Actar-D, 2007), 187.

29 Ibid., 191.

BIBLIOGRAPHY

Addison, Jim. "History of the Mississippi River and Tributaries Project." *US Army Corps of Engineers* New Orleans District. Archived from the original on January 28, 2006. http://wayback.archive.org/web/20060128111022/www.mvn.usace.army.mil/pao/bro/misstrib.htm.

Allen, Stan. "Infrastructural Urbanism." *Cambridge Journal of Architecture: Scroope* 9, 1997.

Allen, Stan. *Practice: Architecture, Techniques and Presentation (revised and expanded edition)*, 251. London/New York: Routledge, 2008.

Aerial Photography Field Office. "Imagery Programs: National Agriculture Imagery Program." *United States Department of Agriculture*, www.fsa.usda.gov/programs-and-services/aerial-photography/imagery-programs/naip-imagery/index June 10, 2014.

Amidon, Jane. "Big Nature." In *Design Ecologies: Essays on the Nature of Design*, edited by Lisa Tilder and Beth Blostein, 165–181. New York: Princeton Architectural Press, 2010.

Architecture as a Technicolor Canvas | 'Emergence' by Obscura. *Vice Creator's Project*, 2013. Film. https://www.youtube.com/watch?v=N2ZMH1XlxBI July, 2015.

Balée, William. "The Research Program of Historical Ecology." *Annual Review of Anthropology* 35 (2006): 75–98.

Balée, William. "People of the Fallow: A Historical Ecology of Foraging in Lowland South America." In *Conservation of Neotropical Forests: Working from Traditional Resource Use*, edited by Kent H. Redford and Christine Padoch, 35–57. New York: Columbia University Press, 1992.

Beesley, Philip. "Introduction: Liminal Responsive Architecture." In *Hylozoic Ground: Liminal Responsive Architecture: Philip Beesley*, edited by Philip Beesley, Hayley Isaacs, Pernilla Ohrstedt, and Rob Gorbet, 12–33. Cambridge: Riverside Architectural Press, 2010.

Beesley, Philip. "Dissipative Models: Notes Toward a Design Method." In *Paradigms in Computing: Making, Machines, and Models for Design Agency in Architecture*, edited by David Jason Gerber and Mariana Ibañez, 24–34. Los Angeles, CA: eVolo Press, 2014.

Beesley, Philip and Omar Kahn. "Responsive Architecture/Performing Instruments." *Situated Technologies, Pamphlet 4: Responsive*. New York: The Architectural League of New York, 2009.

Bennett, Jane. *Vibrant Matter: A Political Ecology of Things*. London: Duke University Press, 2010.

Berger, Alan. *Reclaiming the American West*. New York: Princeton Architectural Press, 2002.

Blum, Andrew. *Tubes: A Journey to the Center of the Internet*. New York: Ecco, 2012.

Blum, Avrim and Tom M. Mitchell. "Combining Labeled and Unlabeled Data with Co-training." Proceedings of the Eleventh Annual Conference on Computational Learning Theory, *COLT*, 92–100. New York: ACM, 1998, doi:10.1145/279943.279962.

Bostrom, Nick. "A History of Transhumanist Thought." *Journal of Evolution and Technology* 14.1 (January, 2005). www.nickbostrom.com/papers/history.pdf. Accessed February 21, 2006.

Brown, Case Lance and Rob Holmes. "Landscape Switching: A New Speed and Territory for Design Agency." *Kerb* 22 (2014): 44–49.

Bryant, Levi R. Onto-Cartography: *An Ontology of Machines and Media*. Edinburgh: University Press, 2014. Accessed July, 2015. www.euppublishing.com/userimages/ContentEditor/1396275575603/Onto-Cartography%20-%20Author%20Q%26A.pdf.

Bullivant, Lucy. "Introduction." *Architectural Design* 75.1 (2005): 5–7.

Bullivant, Lucy. *Responsive Environments: Architecture, Art and Design*. London: V & A Publications, 2006.

Bullivant, Lucy. "Alice in Technoland." *Architectural Design* 77. 4 (2007): 6–13.

Bullivant, Lucy. "Algae Farm." *Domus* (blog). Last modified September 16, 2011. www.domusweb.it/en/architecture/2011/09/16/algae-farm.html.

Burke, Anthony. "Redefining Network Paradigms." In *Network Practices: New Strategies in Architecture and Design*, edited by Anthony Burke and Therese Tierney, 54–77. New York: Princeton Architectural Press, 2007.

Cantrell, Bradley. "Synthetic Mudscapes." In *Paradigms in Computing: Making, Machines, and Models for Design Agency in Architecture*, edited by David Jason Gerber and Mariana Ibañez, 357–361. Los Angeles, CA: eVolo Press, 2014.

Cantrell, Bradley and Justine Holzman. "Synthetic Urban Ecologies project statement." *Reactscape* (blog). Fall 2013. Robert Reich School of Landscape Architecture. Accessed July 1, 2015. http://reactscape.visual-logic.com/teaching/synthetic-urban-ecologies/project-2–0-strategy-and-implementation.

Carey, James. "Technology and Ideology: The Case of the Telegraph." In *The New Media Theory Reader*, edited by Robert Hassan and Julian Thomas. New York: Open University Press, 2006. Previously Published in *Communication as Culture*. London: Routledge, 1989.

Clément, Gilles. *Manifeste du Tiers Paysage*. Montreuil: Sujet-Objet, 2004.

Collet, Carole. "Foreword." In *Alive: Advancements in Adaptive Architecture*, edited by Manuel Kretzer, and Ludger Hovestadt. Basel: Birkhauser Va, 2014.

Corner, James. "Not Unlike Life Itself: Landscape Strategy Now." *Harvard Design Magazine* 21 (Fall/Winter 2004): 32–34.

Corner, James. "The Agency of Mapping: Speculation, Critique and Invention." In *Mappings*, edited by Denis Cosgrove, 213–252. London: Reaktion Books, 1999.

Cosgrove, Denis. "The Measures of America." In *Taking Measures Across the American Landscape*, edited by James Corner and Alex S. MacLean, 3–13. New Haven: Yale University Press, 1996.

Czerniak, Julia, Ed.. *CASE: Downsview Park Toronto*. Munich: Prestel, 2001.

Dale, Nell B. and Chip Weems. *Programming and Problem Solving in Java.* Boston: Jones and Bartlett Publishers, 2007.

Daley, Suzanne. "Greek wealth is found in many places, just not on tax forms." *New York Times,* May 1, 2010. New York edition. www.nytimes.com/2010/05/02/world/europe/02evasion.html?_r=0 July, 2015.

Davis, Brian. "Landscapes and Instruments." *Landscape Journal* 32, no. 2 (2003): 161–176.

Davis, Brian. "Anti-Terraforming and Ecosynthesis, Planetary or Otherwise." *faslanyc* (blog). Last modified June 19, 2012. http://faslanyc.blogspot.com/2012/06/anti-terraforming-and-ecosynthesis.html.

Davison, Nicola. "Rivers of Blood: The Dead Pigs Rotting in China's Water Supply." *The Guardian* (blog). March 29, 2013. www.theguardian.com/world/2013/mar/29/dead-pigs-china-water-supply.

de Monchaux, Nicholas. "Local Code: Real Estates." *Architectural Design* 80, no. 3 (2010): 88–93.

de Monchaux, Nicholas. *Spacesuit: Fashioning Apollo*, 176. Cambridge, Mass.: MIT Press, 2011.

de Monchaux, Nicholas. *Local Code: 3,659 Proposals about Data, Design, and the Nature of Cities.* New York: Princeton Architectural Press, 2015.

de Solà-Morales, Ignasi. "Terrain Vague." In *Terrain Vague: Interstices at the Edge of the Pale.* Eds. Manuela Mariana and Patrick Barron, 24–30. New York: Routledge, 2013.

Dewey, John. *The Quest for Certainty: A Study of the Relation of Knowledge and Action.* New York: Minton, Balch and Company, 1929.

Dewey, John. *Logic: The Theory of Inquiry* (Vol. 12: The Later Works, 1925–38). Carbondale: Southern Illinois Press, 1991.

Dubberly, Hugh, Usman Haque, and Paul Pangaro. "ON MODELING What is interaction?: are there different types?" *Interactions* 16, no. 1 (2009): 69–75.

Duggan, Jennifer. "Dead Pigs Floating in Chinese River." *The Guardian* (blog). April 17, 2014. www.theguardian.com/environment/chinas-choice/2014/apr/17/china-water.

Duhigg, Charles. *The Power of Habit: Why We Do What We Do in Life and Business.* New York: Random House, 2012.

Dunne, Anthony. *Hertzian Tales: Electronic Products, Aesthetic Experience and Critical Design.* Cambridge, Mass.: The MIT Press, 2005.

Easterling, Keller. "Fresh Field." In *Coupling: Strategies for Infrastructural Opportunism*, edited by Neeraj Bhatia, 10–13. New York: Princeton Architectural Press, 2011.

Edergton, Harold. *Electronic Flash, Strobe.* New York: McGraw-Hill, 1970.

Eidner, Franziska and Nadin Heinich. *Sensing Space: Technologien für Architekturen der Zukunft = Future Architecture by Technology.* Berlin: Jovis, 2009.

Ellis, Erle. C. and Navin Ramankutty. "Putting People in the Map: Anthropogenic Biomes of the World." *Frontiers in Ecology and the Environment* 6, no. 8 (2008): 439–447.

Ellsworth, Elizabeth and Jamie Kruse, Eds.. *Making the Geologic Now: Responses to Material Conditions of Contemporary Life.* Brooklyn, NY: Punctum Books, 2013.

Ernstson, Henrik, Sander E. van der Leeuw, Charles L. Redman, Douglas J. Meffert, George Davis, Christine Alfsen, Thomas Elmqvist. "Urban Transitions: On Urban Resilience and Human-Dominated Ecosystems." *AMBIO* 39 (2010): 531–545.

Evans, Helen. "Nuage Vert." *Cluster Magazine* 7 May (2008): 30–35.

Evans, Helen and Heiko Hansen. "On the Modification of Man-Made Clouds: The Factory Cloud." *Leonardo* 47, no. 1 (2014): 72–73.

Evans, Helen, Heiko Hansen, and Joey Hagedorn. "Artful Media: Nuage Vert." *IEEE Multimedia* 16, no. 3 (2009): 13–15.

Fletcher, David. "Flood Control Freakology: Los Angeles River Watershed." In *The Infrastructural City: Networked Ecologies in Los Angeles*, edited by Kazys Varnelis, 36–51. New York: Actar, 2009.

Fox, Michael and Miles Kemp. *Interactive Architecture*. New York: Princeton Architectural Press, 2009.

Fuller, Matthew and Usman Haque. "Urban Versioning System 1.0." In *Situated Technologies Pamphlets 2*, edited by Omar Khan, Trebor Scholz, and Mark Shepard. New York: The Architectural League of New York. Spring 2008.

Fulton, Gale. "Landscape Intelligence." *Journal of Landscape Architecture* 31 (April–June 2011): 46–53.

Galloway, Alexander. *Protocol: How Control Exists After Decentralization*. Cambridge, Mass.: MIT Press, 2004.

Galloway, Alexander and Eugene Thacker. "Protocol, Control, and Networks." *Grey Room* 17 (2004): 6–29.

Galloway, Andrew. "When Biology Inspires Architecture: An Interview with Doris Kim Sung." *ArchDaily* (blog), May 14, 2014. www.archdaily.com/?p=505016 July, 2015.

Galloway, Anne. "Intimations of Everyday Life: Ubiquitous computing and the city," *Cultural Studies* 18, no. 2/3 (March/May, 2004): 384–408.

Gandy, Matthew. "Cyborg Urbanization: Complexity and Monstrosity in the Contemporary City." *International Journal of Urban and Regional Research* 29, no. 1 (2005): 26–49. doi:10.1111/j.1468–2427.2005.00568.x.

Gattegno, Nataly and Jason Kelly Johnson. "Datagrove." *Future Cities Lab*, 2012. www.future-cities-lab.net/projects/#/datagrove/ March 1, 2015.

Gattegno, Nataly and Jason Kelly Johnson. "Theater of Lost Species." *Future Cities Lab*, 2012. Accessed May 1, 2015. www.future-cities-lab.net/theater-of-lost-species/.

Geuze, Adriaan and Maarten Buijs. "West 8 Airport Landscape: Schiphol." *Scenario* 4 (Spring 2014). http://scenariojournal.com/article/airport-landscape/.

Giunchiglia, Fausto and Toby Walsh. "A Theory of Abstraction." *Artificial Intelligence* 56 (1992): 323–390.

Green, Nicola. "On the Move: Technology, Mobility, and the Mediation of Social Time and Space." *In The New Media Theory Reader*, edited by Robert Hassan and Julian Thomas, 249–265. New York: Open University Press, 2006. Previously published in *The Information Society* 18, no. 4 (2002): 281–292.

Greenfield, Adam. *Against the Smart City (The City is Here for You to Use Book 1)*. New York City: Do Projects, Pamphlet 1302, October 13, 2013.

Gumprecht, Blake. *The Los Angeles River: Its Life, Death, and Possible Rebirth*. Baltimore: Johns Hopkins University Press, 1999.

Hammond, William A. "Hylozoism: A Chapter in the Early History of Science." *The Philosophical Review* 4, no. 4 (1895): 394–406.

Haque, Usman. "Hardspace, Softspace and the Possibilities of Open Source Architecture." (2002). Accessed July 2015. www.haque.co.uk/papers/hardsp-softsp-open-so-arch.PDF.

Haque, Usman. "Architecture, interaction, systems." *Archiquitetura & Urbanismo* 149 (August 2006).

Haque, Usman. "The Architectural Relevance of Gordon Pask." *Architectural Design* 77, no. 4 (2007): 54–61.

Haque, Usman and Adam Somlai-Fische. "Low Tech Sensors and Actuators for Artists and Architects." In *Research: The Itemisation of Creative Knowledge/ Editor*, edited by Clive Gillman. Liverpool, UK: FACT: Liverpool University Press.

Hedonometer.org. "Average Happiness for Twitter." Accessed July 1, 2015. http://hedonometer.org/index.html.

Hill, Kristina. "Shifting Sites." In *Site Matters: Design Concepts, Histories, and Strategies*, edited by Carol J. Burns and Andrea Kahn, 131–156. New York, NY: Routledge, 2005.

Hockenos, Paul. "Street Smarts: From Holland, Bright Ideas for Highways." *New York Times*. April 28, 2013, New York edition. Accessed July 1, 2015. www.nytimes.com/2013/04/28/automobiles/from-holland-bright-ideas-for-highways.html?_r=1.

Holmes, Rob. "New American Wilderness." Paper presented at 2014 *CELA Annual Conference*. Baltimore, Maryland, March 27, 2014.

Howard, John A. and Colin W. Mitchell. *Phytogeomorphology*. New York: John Wiley & Sons, 1985.

Höweler, Eric and Meejin Yoon. "Wind Screen." *Höweler + Yoon Architecture* (blog). 2011. Accessed July 1, 2015. www.hyarchitecture.com/projects/48.

Höweler, Eric and Meejin Yoon. "Swing Time." *Höweler + Yoon Architecture* (blog). 2014. Accessed July 1, 2015. www.hyarchitecture.com/projects/114.

Jackson, John Brinkerhoff. "The Vernacular Landscape." In *Landscape Meanings and Values*, edited by David Lowenthal and Edmund C. Penning-Rowsell, 65–77. London: Allen and Unwin, 1986.

Jacobs, Jane. *The Death and Life of Great American Cities*. New York: Random House, 1961.

Jereminjenko, Natalie. "Introduction." *The xClinic Environmental Health Clinic*, NYU. Accessed July, 2015. www.environmentalhealthclinic.net/environmental-health-clinic.

Johnson, Jason Kelly. "Thinking Things, Sensing Cities." In *Architecture In Formation: On the Nature of Information in Digital Architecture*, edited by Pablo Lorenzo-Eiroa and Aaron Sprecher. Routledge: New York, 2013.

Kammerer, J.C. "Largest Rivers in the United States." *U.S. Geological Survey*, May, 1990. Last modified September 1, 2015. http://pubs.usgs.gov/of/1987/ofr87-242/.

Kane, Will. "Oakland Cops Aim to Scrap Gunfire-Detecting ShotSpotter." *SFGate* (blog). March 14, 2014. www.sfgate.com/crime/article/Oakland-cops-aim-to-scrap-gunfire-detecting-5316060.php July, 2015.

Kwinter, Sanford and Cynthia C. Davidson. *Far from Equilibrium: Essays on Technology and Design Culture*. Barcelona: Actar-D, 2007.

LeCavalier, Jesse. "All Those Numbers: Logistics, Territory and Walmart." *Places Journal*, May, 2010. https://placesjournal.org/article/all-those-numbers-logistics-territory-and-walmart/.

Lee, Heejin and Jonathan Lievenau. "Time and the Internet." In *Time and Society* 9, no.1 (2001): 48–55. Reprinted in The New Media Theory Reader, edited by Robert Hassan and Julian Thomas. New York: Open University Press. 2006.

Leopold, Donald. "Distortion of Olfactory Perception: Diagnosis and Treatment." *Chemical Senses* 27, no. 7 (2002): 611–615.

Ley, Rob and Joshua G. Stein. "Reef." *Rob Ley* (blog). 2009. Accessed July 1, 2015. http://rob-ley.com/Reef.

Lister, Nina-Marie. "Ecological Design or Designer Ecologies?" In *Large Parks*, edited by Julia Czerniak and George Hargreaves. New York: Princeton Architectural Press, 2007.

Lister, Nina-Marie and Chris Reed. "Introduction: Ecological Thinking, Design Practices." In *Projective Ecologies*, edited by Chris Reed and Nina-Marie Lister, 14–21. New York, NY: Actar, Harvard Graduate School of Design, 2014.

Makarieva, Anastassia. "Cybernetics." In *Systems Ecology: Encyclopedia of Ecology* (Volume 1), edited by Sven Erik Jørgensen and Brian D. Fath, 806–812. Oxford: Elsevier, 2008.

Manaugh, Geoff, Ed.. *Landscape Futures: Instruments, Devices and Architectural Inventions*. Barcelona: Actar, 2013.

Manovich, Lev. *Software Takes Command (Vol. 5): International Texts in Critical Media Aesthetics Founding*, edited by Francisco J. Ricardo, New York, NY: Bloomsbury, 2013.

Marris, Emma. *Rambunctious Garden: Saving Nature in a Post-wild World*. New York, NY: Bloomsbury, 2011.

Matsuda, Keiichi. "Domesti/City: The Dislocated Home in Augmented Space." *Master of Architecture Thesis*, University College London, Bartlett, 2010.

Matta-Clark, Gordon, Jeffrey Kastner, Sina Najafi, Frances Richard, and Jeffrey A. Kroessler. *Odd Lots: Revisiting Gordon Matta-Clark's "Fake Estates"*. New York: Cabinet Books and Queens Museum of Art and White Columns, 2005.

Mattern, Shannon. "Methodolatry and the Art of Measure." *Places Journal*, November, 2013. https://placesjournal.org/article/methodolatry-and-the-art-of-measure/.

May, John. "Logic of the Managerial Surface." *Praxis* 13 (2011): 116–124.

McCullough, Malcolm. *Abstracting Craft: The Practiced Digital Hand*. Cambridge, Mass.: MIT Press, 1996.

McCullough, Malcolm. *Digital Ground: Architecture, Pervasive Computing, and Environmental Knowing*. Cambridge, Mass.: MIT Press, 2004.

McHarg, Ian L. *Design with Nature*. Garden City, NY: Natural History Press, 1969.

McLuhan, Marshall. "The Relation of the Environment to the Anti-Environment." In *Marshall McLuhan—Unbound*, 4th edn Corte Madera, CA: Ginko Press, 2005.

Medford, June I. and Ashok Prasad. "Plant Synthetic Biology Takes Root: Applying the Basic Principles of Synthetic Biology to Plants Shows Progress." *Science* 346, no. 6206 (2014): 162–163.

Millet, Lydia. "The Child's Menagerie." Opinion Editorial, *New York Times*, December 9, 2012. New York edition.

Moggridge, Bill. *Designing Interactions*. Cambridge, Mass.: The MIT Press, 2007.

Morton, Timothy. *The Ecological Thought*. Cambridge, Mass.: Harvard University Press, 2010.

Morton, Timothy. "Zero Landscapes in the Time of Hyperobjects." *Graz Architecture Magazine* 07 (2011): 78–87.

Mostafavi, Mohsen and Gareth Doherty, Eds., *Ecological Urbanism*. Baden, Sweden: Lars Müller, 2010.

Nye, David E. *American Technological Sublime*. Cambridge, Mass.: MIT Press, 1994.

Odling-Smee, F. John, Kevin N. Laland and Marcus W. Feldman. *Niche Construction: The Neglected Process in Evolution*. Princeton, NJ: Princeton University Press, 2003.

Orff, Kate. "Re-Reading: Rachel Carson, 'Undersea' (1937)." *Harvard Design Magazine* 39 (Fall/Winter 2014): 115.

Pask, Gordon. "cybernetics." *Encyclopædia Britannica*, 14th edn, 1972.

Pask, Gordon. *Conversation Theory: Applications in Education and Epistemology*. Amsterdam and New York: Elsevier Publishing Co., 1976.

Pasquero, Claudia and Marco Poletto. "HORTUS." *ecoLogic Studio* (blog). January 13, 2012. www.ecologicstudio.com/v2/project.php?idcat=3&idsubcat=59&idproj=115.

Pasquero, Claudia and Marco Poletto. "METAfolly." *ecoLogic Studio* (blog), March 2, 2013. www.ecologicstudio.com/v2/project.php?idcat=3&idsubcat=66&idproj=120.

Pickett, S. T. A., M. L. Cadenasso, 1 J. M. Grove, C. H. Nilon, R. V. Pouyat, W. C. Zipperer, and R. Costanza. "Urban Ecological Systems: Linking Terrestrial Ecological, Physical and Socioeconomic Components of Metropolitan Areas." *Annual Review Ecological Systems* 32 (2001): 127–157.

Pousman, Zachary and John Stasko. "A Taxonomy of Ambient Information Systems: Four Patterns of Design", 67–74. *Proceedings of the Working Conference on Advanced Visual Interfaces*. New York, NY: ACM, 2006.

Rabinow, Paul. *Marking Time: On the Anthropology of the Contemporary*. Princeton, NJ: Princeton University Press, 2008.

Raxworthy, Julian. "Novelty in the Entropic Landscape: Landscape Architecture, Gardening, and Change." *Doctoral Dissertation*, University of Queensland, 2013. (eSpace) www.academia.edu/8879242/Raxworthy_Julian._2013._Novelty_in_the_Entropic_Landscape_Landscape_architecture_gardening_and_change_School_of_Architecture_University_of_Queensland_Brisbane.

Reed, Chris. "The Agency of Ecology." In *Ecological Urbanism*, edited by Mohsen Mostafavi, and Gareth Doherty. Baden, Sweden: Lars Müller, 2010.

Reed, Chris and Nina-Marie Lister. "Parallel Genealogies." In *Projective Ecologies*, edited by Chris Reed and Nina-Marie Lister. New York, NY: Actar, Harvard Graduate School of Design, 2014.

Robinson, Alexander. "Owens Lake Rapid Landscape Prototyping Machine: Reverse-Engineering Design Agency for Landscape Infrastructures." In *Paradigms in Computing: Making, Machines, and Models for Design Agency in Architecture*, edited by David Jason Gerber and Mariana Ibañez, 348–356. Los Angeles, CA: eVolo Press, 2014.

Ross, Andrew. "The New Smartness." In *Culture on the Brink: Ideologies of Technology*, edited by Gretchen Bender and Timothy Druckrey. Seattle: Bay Press, 1994. Series: Discussion in Contemporary Culture, no. 9.

Satellite Sentinel Project. "Documenting the Crisis." Accessed April 10, 2015. www.satsentinel.org/documenting-the-crisis.

Schlosser, Eric. *Fast Food Nation: The Dark Side of the All-American Meal*. Boston: Houghton Mifflin, 2001.

Schüll, Natasha Dow. "The Folly of Technological Solutionism: An Interview with Evgeny Morozov." *Public Books* (blog). September 9, 2003. www.publicbooks.org/interviews/the-folly-of-technological-solutionism-an-interview-with-evgeny-morozov.

Scott, James C. *Seeing Like a State: How Certain Schemes to Improve the Human Condition Have Failed*. New Haven, CT, USA: Yale University Press, 1998.

Strang, Gary. "Infrastructure as Landscape." *Places* 10, no. 3 (1996): 8–15.

Sung, Doris. "Bloom." *dO|Su Studio Architecture* (blog). August, 2012. http://dosu-arch.com/bloom.html.

Sutton, Richard S. and Andrew G. Barto. *Reinforcement Learning: An Introduction*. Cambridge, Mass.: MIT Press, 1998.

Trancoso, Ralph, Edson E. Sano, and Paulo R. Meneses. "The Spectral Changes of Deforestation in the Brazilian Tropical Savanna." *Environmental Monitoring and Assessment: An International Journal Devoted to Progress in the Use of Monitoring Data in Assessing Environmental Risks to Man and the Environment* 187, no. 1 (2015): 1–15.

Turing, Alan. "Computing Machinery and Intelligence." *Mind: A Quarterly Review of Psychology and Philosophy* 59 (January 1950): 433.

United States Census Bureau. "About the Bureau." Accessed July 1, 2015. *United States Census Bureau* www.census.gov/about/what.html.

Valdivia y Alvarado, Pablo, Vignesh Subramaniam, and Michael Triantafyllou. "Design of a Bio-inspired Whisker Sensory for Underwater Applications." *Conference: Sensors*, 28–31 Oct, 2012, doi:10.1109/ICSENS.2012.6411517.

van Loon, Joost. "Social Spatialization and Everyday Life." *Space and Culture* 5, no. 2 (2002): 88–95.

Varnelis, Kazys, Ed.. *The Infrastructural City: Networked Ecologies in Los Angeles*. Barcelona: Actar, 2009, 118–129.

Waldheim, Charles. *The Landscape Urbanism Reader*. New York: Princeton Architectural Press, 2006.

Waldheim, Charles. "Strategies of Indeterminacy in Recent Landscape Practice." *Public* 33 (Spring 2006): 80–86.

Wark, McKenzie. "Abstraction/class." In *The New Media Theory Reader*, edited by Robert Hassan and Julian Thomas, 212–219. New York: Open University Press, 2006.

Weibel, Peter. "Intelligent Ambience." *Promotional leaflet for Ars Electronica*. Linz: ORF, 1994.

Weintraub, Linda. *To Life!: Eco art in pursuit of a sustainable planet*. Berkeley, CA: University of California Press, 2012.

Weiser, Mark. "The Computer for the 21st Century." *Scientific American* 265 no. 3 (1991): 94–104.

Weiser, Mark and John Seely Brown. "The Coming Age of Calm Technology." *Xerox PARC*, 1996. Accessed July 1, 2015. www.ubiq.com/hypertext/weiser/acmfuture2endnote.htm.

Yoon, J. Meejin and Eric Höweler. *Expanded Practice: Höweler + Yoon Architecture/My Studio*. New York: Papress, 2009.

INDEX